BIZARRE PHENOMENA

QUEST FOR THE UNKNOWN

BIZARRE
PHENOMENA

THE READER'S DIGEST ASSOCIATION, INC.

Pleasantville, New York/Montreal

Quest for the Unknown
Created, edited, and designed by DK Direct Limited

A DORLING KINDERSLEY BOOK

DK DIRECT LIMITED

Series Editor Richard Williams
Editors Deirdre Headon, Tony Whitehorn
Editorial Research Julie Whitaker

Senior Art Editor Susie Breen
Art Editor Mark Osborne
Designer Juliette Norsworthy
Senior Picture Researcher Frances Vargo; **Picture Assistant** Sharon Southren

Editorial Director Jonathan Reed; **Design Director** Ed Day
Production Manager Ian Paton

Volume Consultant Peter Brookesmith
Contributors Larry E. Arnold, Chris Cooper, Melvin Harris, Michael A. Hoffman II,
Robert Kiener, Prof. G. T. Meaden, Dr. J. T. Richards, Paul Sieveking, Val Stevenson

Illustrators Ian Craig, Roy Flooks, Richard Manning, Emma Parker,
Darrel Rees, Matthew Richardson, Mike Shepherd
Photographers Zafer Baran, Simon Farnhell, Andrew Garner,
Mark Hamilton, Gary Marsh, Susanna Price, Paul Venning, Alex Wilson

Library of Congress Cataloging in Publication Data

Bizarre phenomena
     p.   cm. — (Quest for the unknown)
   "A Dorling Kindersley book" — T.p. verso.
   Includes index.
   ISBN 0-89577-464-X
   1. Science—Miscellanea.   I. Reader's Digest Association.
   II. Series.
   Q173.B624 1992
   500—dc20   92-31739

Printed in the United States of America

FOREWORD

*T*HE UNUSUAL SEEMS TO HAPPEN FREQUENTLY, and the impossible is slightly rarer. Such is the conclusion to which many of us might be led by the examples of bizarre phenomena in this volume. Because, throughout the ages and in every part of the world, the strange and yet sometimes wonderful has been recorded regularly and in great detail: live toads entombed in rocks, human beings with magnetic powers, the appearance of giants and dwarfs, and coincidences so unusual as to seem to be beyond the power of the human imagination to invent.

Undoubtedly, many of these phenomena are imaginary, fake, or fraudulent; some, when closely investigated, have turned out to have a perfectly rational cause. Yet the sheer number of puzzling cases has led some commentators to question the efficacy of the scientific approach. For it seems that science has never been, and may never be, able to account for everything that happens on our planet.

In 1772 the residents of Luce in France reported that a large rock had fallen from the sky, causing considerable damage. During the resulting investigation by the Académie Française, the respected chemist Antoine Lavoisier pronounced with absolute certainty that the stone could not have fallen from the sky, because there *are* no stones in the sky. The strength of Lavoisier's conviction was such that it helped put back the scientific study of meteorites for over a quarter of a century.

Let us beware, therefore, of taking too narrow a view of bizarre phenomena. Many such reports may be difficult to explain or believe. But just because researchers cannot yet account for all such reported incidents, it is important to be wary of ignoring what evidence there might be and stubbornly declaring that such things cannot happen.

— The Editors

CONTENTS

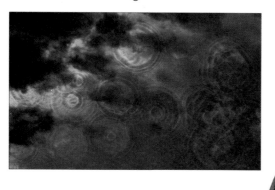

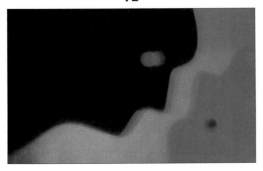
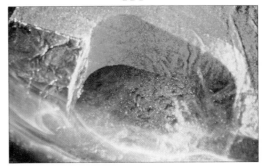

MYSTERIES IN OUR MIDST

Frogs falling like rain, severe burns occurring with no apparent heat source, the image of a dead girl appearing spontaneously on a billboard — these strange events are just a foretaste of the many bizarre phenomena to which we introduce you in this volume.

Over the years many stories have been told of showers of animals descending from the sky. Frogs, fish, newts, crabs, and birds have all been reported as having fallen — and the scientific world has generally reacted with skepticism. This account, from England, of a young farm girl experiencing a bizarre thunderstorm and a sudden rain of living frogs, is typical.

"When I was about 10 years old (I'm now 62) and living at Rode (Somerset)...it was the custom of my father, after milking the

IDENTICAL DEATHS

In Hamilton, Bermuda, in July 1974, 17-year-old Neville Ebbin was knocked off his moped by a cab and killed. The following year another young man was killed in exactly the same way. The victim was Neville Ebbin's brother, Erskine. He was the same age as Neville when he was killed. The month was the same. He was riding the same moped. The accident occurred in the same street. It was the same cab that had killed his brother. The same driver was at the wheel. Inside the cab was the same passenger.
(Source: Buenos Aires *Herald*, July 22, 1975)

LOST AND FOUND

On January 14, 1992, Scottish fisherman Jim Gault was lost overboard from the fishing boat Dayspring *40 miles off the northeast coast of Scotland. An air and sea search failed to recover his body. Three months later, on April 7, while his brother William was fishing from the same boat, he hauled in a human body in his net. From the clothes on the corpse, he recognized it as his brother.*
(Source: London *Evening Standard*, April 8, 1992)

cows, to send the beasts down to the pasture under the care of our trained dogs; I would then have to go down and close the gate once they were all in.

"On this particular occasion the animals were pastured in the 'brickyard field,' which adjoined an old brickyard long fallen out of use and which, for many years, had become a large pond of fathomless depths, swamp-edged and covered with reeds and water lilies. It must have been about the turn of the month of July/August for I had not been at school.

"The sky had become a very strange (to me) milky green, rather ominous, and seemed to presage a storm — not that that bothered me for, unlike my mother who was terrified of such things, I rather enjoyed a good thunderstorm! Later a knowledgeable neighbor told me that we had experienced an electric storm.

Rain of frogs

"As I went down the hill toward the field gate, the old dog (who always waited until someone arrived to close the gate) began to contort himself as if having a fit, and suddenly I felt the splat of what I at first thought to be heavy raindrops. Very soon I realised that what was falling was not rain but myriads of tiny frogs! I quickly pulled my blouse up over my head and ran toward the gate, thinking I suppose, to escape them. The dog was still going berserk because, being a

> ## "Suddenly I felt the splat of what I at first thought to be heavy raindrops. Very soon I realised that what was falling was not rain but myriads of tiny frogs!"
> **Mrs. Mabs Holland**

collie, the frogs were in his hair. The 'storm' stopped as suddenly as it had first begun but the grass was literally alive, just for little more than seconds.

Repugnance

"My initial repugnance at having frogs in my hair soon faded, although at first I thought I was responsible for some dire 'plague of frogs' (being rather religiously brought up, then). The dog soon became normal, and I suppose the movement had stopped on his old body. Then I looked closer at some of the tiny little creatures and found they were dying almost instantaneously.

Washed away

"Then the storm broke. I ran home to tell my mother about the frogs. She gave me a good hiding for telling fibs! Father, when he came home from work that evening, seemed to believe me and said he would look the following morning when he went to get the cows. He found nothing but, as he said, birds could have eaten them — apart from which the grass was very long and the gutters had run with water after the storm and washed them away — or perhaps they had just disintegrated?"

Account by Mrs. Mabs Holland, Shepton Mallet, England, published in **The Journal of Meteorology *(U.K.), October 1986.***

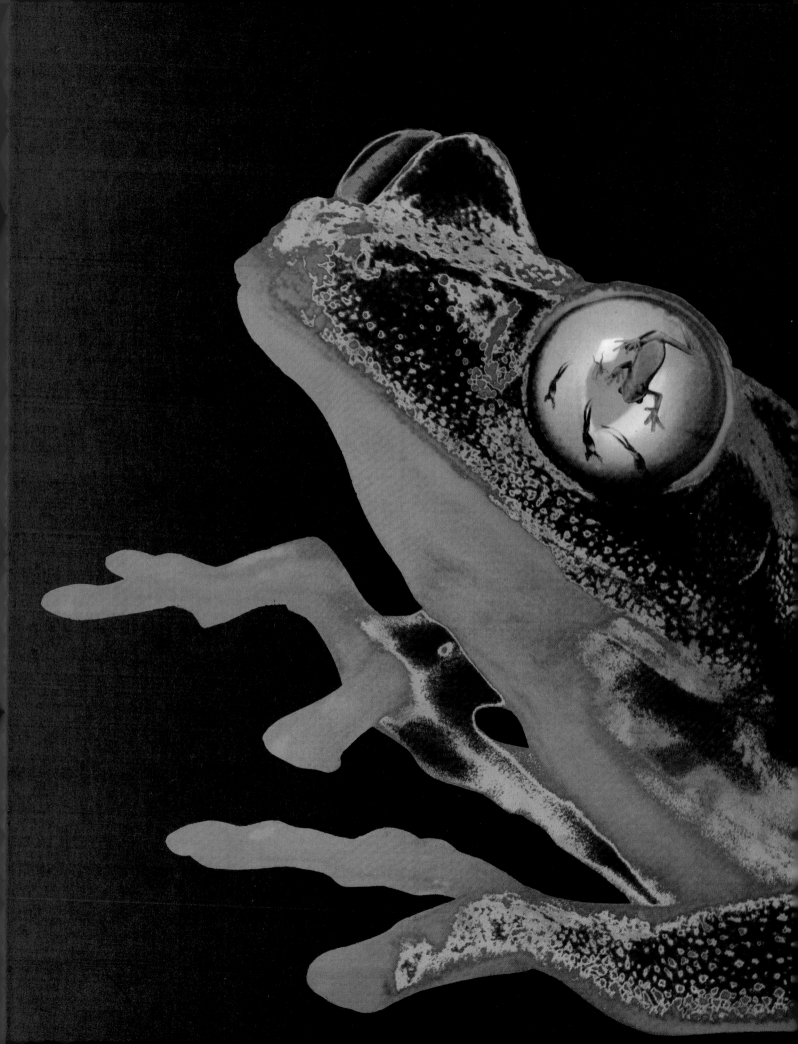

PROFIT AND LOSS

One morning in 1990 Steve Mallinson, of Rickmansworth, England, who had won a sports forecasting competition, eagerly opened the letter containing his prize check. It was for £371.70. In the same post was a letter from the tax man. It was a bill for £371.70.
(Source: London *Daily Mirror*, October 2, 1990)

BURNING WITH RAGE

Throughout history there have been numerous reports of an extraordinary phenomenon known as spontaneous human combustion — incidents of people allegedly burning of their own accord, without any apparent form of external ignition. Here is one such account, published for the first time.

Dr. Sullivan's story

At 6 o'clock on the evening of April 24, 1970, Dr. E. J. Sullivan, a physician and retired U.S. Navy commander, collapsed on the bed in his room at the Vernon Manor Hotel in Cincinnati, Ohio. He had just finished making arrangements for the funeral of his wife, who had died that morning. Now, filled with grief and impotent rage at having lost the woman he loved, he had reached what he felt was the lowest point of his life.

"My next memory," said Dr. Sullivan later, "is becoming aware of a phone ringing, and trying to fight my way up out of whatever I was in to answer it. I gradually came to realize something was wrong. I tried to get up, but could not.

"I tried to open my hands. They would not open. My left hand was all blotchy.... The blotches were blisters. One great amber fluid-filled blister ran along the side of my frozen thumb.

"Had I been in a fire? Burns should hurt, but my left forearm didn't. The hand itself was numb, as free of feeling as though it belonged to someone else, and was mine no more....I lay there, trying to figure out what had happened."

Blistered arms and legs

Dr. Sullivan's cries alerted a hotel employee, who summoned help. George M. Lawton, Dr. Sullivan's personal physician, arrived and examined his patient. He found that blisters covered Sullivan's hands, forearms, legs, and feet. Despite Sullivan's protests that he had not been in contact with anything electrical, and despite the fact that there was not so much as a scorch mark on his bed, Lawton diagnosed the cause of the blisters as electrical burns and said that Sullivan required admission to the hospital.

Lost hours

Then the mystery took a fresh turn. While waiting for an ambulance to take him to the nearby Cincinnati Jewish Hospital, Dr. Sullivan saw that the time was now 6 P.M. on April 25. Twenty-four hours had elapsed since he had originally stretched out on his bed — and he could account for only a few of them!

The burns had caused Dr. Sullivan's hands to lose their ability to grip — a loss that turned out to be permanent. Dr. Lawton's diagnosis of electrical burns was finally rejected, but each alternative theory was also discarded.

Haunted by the mystery

The mystery of what had happened to him during those lost 24 hours haunted Dr. Sullivan ceaselessly. Nothing in his medical training could begin to suggest a solution to him. Then one day he heard about spontaneous human combustion, and suddenly he was convinced that he had discovered the source of his burns. "If ever there was an example of 'burning with rage,' I personified it that day. I had lost...everything that meant anything to me at all, and I was a container of rage."

Dr. Sullivan believed that somehow, in some strange manner beyond present scientific understanding, the fierce heat of his emotions had translated itself into an actual physical heat within, so powerful that it had caused his body to ignite....

Account by Dr. E. J. Sullivan given exclusively to Larry E. Arnold, director of ParaScience International.

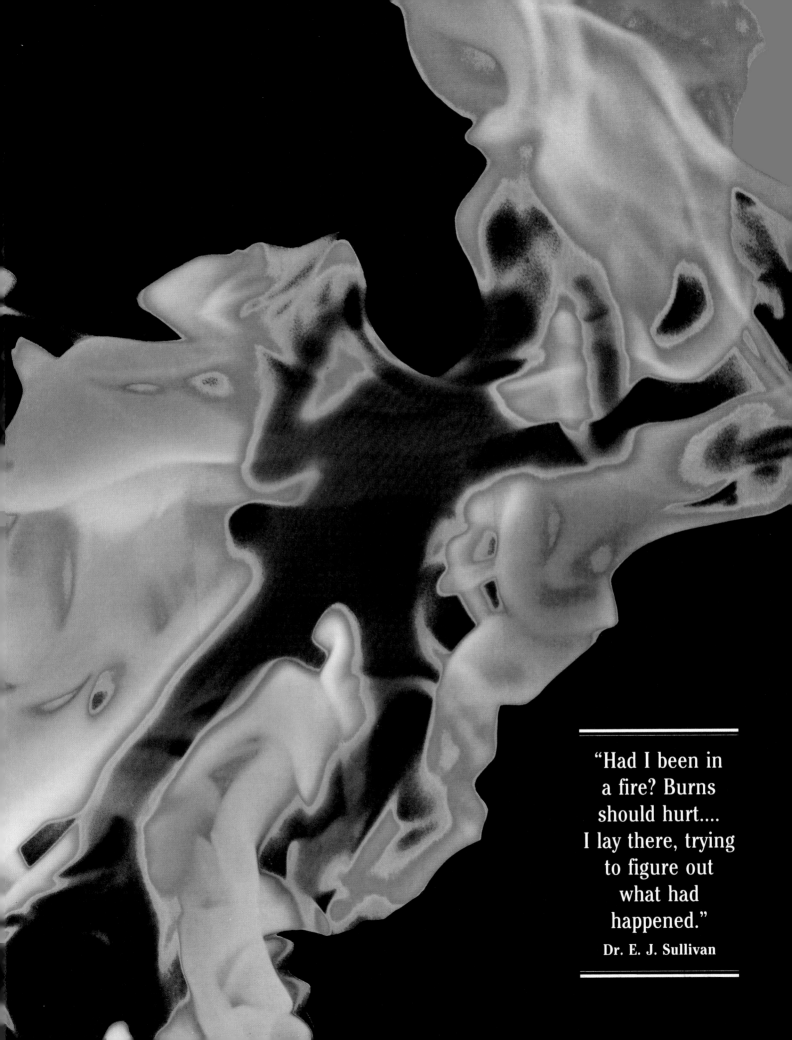

"Had I been in
a fire? Burns
should hurt....
I lay there, trying
to figure out
what had
happened."

Dr. E. J. Sullivan

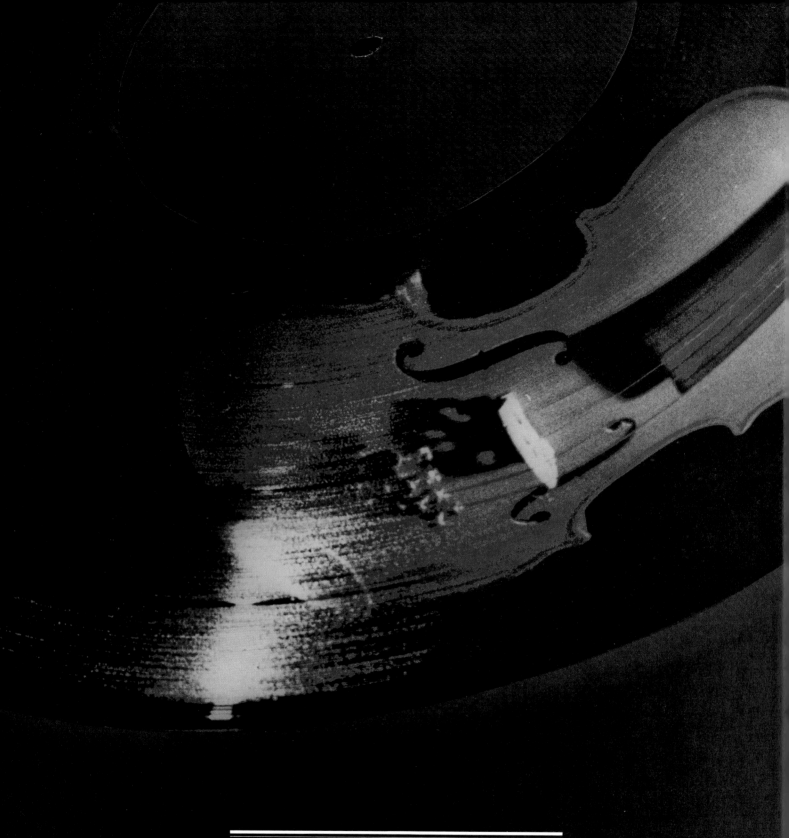

"Arthur Lintgen apparently possesses the remarkable ability to identify the music on a phonographic record without referring to its label."
James Randi

PLUM LUCKY

While at school the 19th-century French poet Emile Deschamps was given a taste for English plum pudding by a M. de Fortgibu.

Ten (plum-pudding-less) years later, Deschamps entered a restaurant and asked for a piece of plum pudding he had seen being prepared. Unfortunately, it was reserved for another diner. This turned out to be M. de Fortgibu. After an interval of many (again plum-pudding-less) years, Deschamps was at a dinner party where plum pudding was to be served.

During the meal someone who had been invited to dinner at another apartment in the building but had lost his way turned up in error at Deschamps' hosts' apartment. It was M. de Fortgibu. "Three times in my life have I eaten plum pudding," exclaimed Deschamps, "and three times have I seen M. de Fortgibu!" (Source: Camille Flammarion, *The Unknown*, 1902)

RECORD ACHIEVEMENT

In this volume we shall encounter a number of strange people with seemingly extraordinary powers and talents. Some appear to be generators of electricity, some of magnetism, and some are impervious to heat. The following account — from one of the most skeptical sources possible — tells of a talent that may be unique in the annals of strange phenomena.

Dr. Lintgen's amazing talent

Arthur Lintgen, a Pennsylvania physician, apparently possesses the remarkable ability to identify the music on a phonographic record without referring to its label. He identifies it simply by looking at the grooves on the surface of the disc. Dr. Lintgen imposes only one restriction on challenges to his strange

talent — the music must be classical, orchestral, and post-Mozart. Even with such restrictions, this leaves a body of recorded music that runs into thousands of records.

Covered labels

James Randi, a professional magician and well-known investigator of seemingly paranormal events, was commissioned by the magazine *Discover* to test the doctor's strange ability. A skeptical Randi collected some recordings, covered the labels and matrix numbers on them, and visited Lintgen. The doctor's compelling demonstration convinced and astounded the hard-headed investigator.

"Lintgen immediately spotted two different recordings of Stravinsky's *Rite of Spring*," wrote Randi, and "correctly identified...Ravel's *Bolero*, Tchaikovsky's *1812 Overture*, Holst's *The Planets*, and Beethoven's Sixth Symphony — though he expressed puzzlement over an extra cut on the disc, finally identifying it as Beethoven's *Prometheus Overture*."

Magician

Randi also presented Lintgen with "control" items. The doctor declared one to be "gibberish" and another to be "a vocal solo of some kind." The first turned out to be a recording by rock star Alice Cooper, and the second to be a recorded lecture called "So You Want to Be a Magician." "Certainly," concluded Randi, "Arthur Lintgen comes as close to that definition as I ever hope to see!"

Account published in **The Skeptical Inquirer, Summer 1982.**

RETURN VISIT

In 1966 Cristina Corta, of the city of Salvador, Brazil, had a distressing experience. A truck ran out of control and crashed into her house. In 1989 Corta was in the house when a thunderous noise and falling plaster brought her to her front door. Her property had been virtually demolished.

"Not you again," she exclaimed unbelievingly. Before her stood the selfsame driver responsible for the first crash 23 years before.

(Source: *Sandwell Express & Star*, England, November 8, 1989)

BLIND FORTUNE

On April 23, 1990, in Fernandina Beach, Florida, Mrs. Margaret Waldon, a 74-year-old blind woman, amazed her golfing partners. At the seventh hole she made a hole in one. They were even more amazed the following day. On the same green Mrs. Waldon holed in one again.

(Source: London *Daily Telegraph*, April 27, 1990)

POSTHUMOUS PORTRAIT

From time to time images of faces, words, and numbers, among other forms, are reported to have appeared spontaneously in the most unlikely places. Some who witness them see no more than chance patterns devoid of meaning — but others believe that they are of supernatural origin and that, in some cases, they convey a message. Whether or not such phenomena exist only in the eye of the beholder, there are those that have had the strange power to convince crowds of people that they are meaningful. The sad story that follows is about one such image.

Looking for Laura

On June 19, 1991, nine-year-old schoolgirl Laura Arroyo was abducted from her San Diego home in California and murdered. Her body was found at an industrial park in Chula Vista, to the south. In early July a strange sight began appearing on a blank white billboard about two miles away from the Arroyos' home and from the spot where Laura's body was found. In the evening, as the light faded from the sky and the scene was lit by street lamps, there appeared on the blank billboard what seemed to be an image of the dead girl, gazing toward the sky.

> As the light faded from the sky and the scene was lit by street lamps, there appeared on the blank billboard what seemed to be an image of the dead girl, gazing toward the sky.

Every evening crowds began to gather to watch the gradual, spontaneous appearance of the posthumous portrait. Luis Arroyo, Laura's father, drove his wife and two young sons to see the image. He did not tell the boys the purpose of their journey. Yet, he said, the boys saw the billboard picture of their murdered sister for themselves as the automobile drew near, and started screaming: "Laura! Laura!"

Killer's face

On the night of July 18 the crowd at the site had swelled to between 10,000 and 25,000, and all traffic had come to a standstill. A week later the billboard company switched off the lights illuminating the billboard, in an effort to restore peace to the area. However, residents reconnected them. Kelly Harmon, a resident of Chula Vista, claimed that the billboard showed not only the dead girl but her killer. "Off to the right you can see his face, too," she said. "Eventually, her face will fade away, and all we'll see is his face. That's when he'll be caught."

Account published in **The New York Times, July 28, 1991.**

THE FORTEAN UNIVERSE

Fascinated by worldwide reports of strange phenomena, an American named Charles Fort began collecting them. The highly individual books he filled with such data presented a new way of looking at the universe and helped lay the foundations for today's widespread interest in the paranormal.

*S*TRANGE PHENOMENA that seem to defy rational explanation, like those just described at the beginning of this chapter, have been reported in the thousands throughout history. Such seemingly paranormal happenings are known today, by those who study them, as Fortean phenomena, after the pioneering American anthologist of the bizarre, Charles Fort (1874–1932).

As a young man, Fort began jotting down, from a multitude of publications, accounts of exceptional occurrences of all kinds. He noticed that scientists working in the areas in which these anomalies had reportedly taken place — such as naturalists, geologists, and physicists — tended to ignore, suppress, or discredit all such stories. It appeared they did this because these accounts did not fit in with their fixed ideas of what was possible and what was not.

"The damned"

As a result, Fort believed that science was failing in its duty to examine data of all kinds — even the seemingly outlandish — with an open mind. In arriving at its conclusions, scientific thought of his day seemed to him to be determined to exclude rather than include as much data as possible. He believed it to be dominated, like many organized religions, by orthodoxy and dogma — and, indeed, he came to regard science as simply another form of religion, "excommunicating" facts, of the kind collected by Fort himself, that challenged accepted scientific opinion. Such "excommunicated" data Fort therefore called, perhaps rather fancifully, "the damned,"

POINTS OF VIEW

There are three possible attitudes toward the published reports of bizarre phenomena. One is to be utterly skeptical and dismiss all accounts of the apparently paranormal as worthless — the product of delusions, overheated imaginations, or hoaxes. The numbers of witnesses to many seemingly supernatural incidents, however, and the respectability of many observers, surely suggest that such automatic dismissal may not be sensible.

Frauds

Another response is that since the incidents are attested to and in print, they are very likely true. But in view of so many exposed frauds in the field of claimed paranormal happenings, and in view of the undoubted unreliability of some witnesses, this attitude may be equally ill-founded.

An open mind

A third approach, and that adopted by Fort, may be the most sensible: to maintain an open mind toward all reports of the exceptional, and even the downright unbelievable. It would be arrogant to believe that man had already discovered all the energies in the universe. Could some bizarre phenomena, of the kind you will read about in this book, be due to forces presently beyond our comprehension? Surely we should not reject such a notion out of hand.

and he called his first collection of these bizarre phenomena, published in New York in 1919, *The Book of the Damned*.

Rain of blood

The Book of the Damned was a most impressive, and compulsive, compilation of hundreds of reports of mysterious and unexplained incidents from all over the world, complete with a reputable published source for each one. For example, Fort quoted stories of blood falling from the sky in Calabria, Italy, in 1890; of vast quantities of fish scales found along a 40-mile stretch of the Mississippi River's banks in 1873; and of varicolored lights moving at high velocity over the coast of Wales in 1877.

But Fort was not necessarily arguing for the existence of the supernatural. He was perfectly prepared to accept that there might be a rational explanation for every one of the phenomena he had listed. All he wanted was that these reported happenings should be looked at with as open a mind as that brought to any other type of event.

"Underlying oneness"

One key element of Fort's reasoning was his emphasis on the "underlying oneness" or "continuity" of existence. He suggested that we should see *everything* in the universe as related. For example, coral islands, though apparently separate and very different from one another, are in fact projections from a common substance on the ocean bed.

On this matter, Fort once again took issue with science. Its constant thrust, he claimed, was away from unity and toward breaking down existence into ever smaller subdivisions and slotting together whichever phenomena it did accept into watertight categories.

In *The Book of the Damned* Fort attempted to show the falsity of the rigidly categorizing approach by the following analogy. Redness and yellowness were, he said, forms of exact classification as reasonable as any other that science had formulated. Let us imagine therefore, he hypothesized, that science should take all things red as true and all things yellow as illusory. In that case, Fort argued, "the demarcation would have to be false and arbitrary, because things colored orange, constituting continuity, would belong on both sides of the attempted borderline."

The implication of Fort's example was that in the realm of strange phenomena, there exists, somewhere between the world of incontrovertibly verifiable data and the shadowy domain of myth, delusion, and hoax, a twilight world of phenomena that cannot be said to fit into either of the other two categories. Fort believed that this domain demanded serious investigation, and that it might then yield some startling enhancement of our present knowledge of the world.

Teleportation

Another idea that Fort put forward for consideration in *The Book of the Damned* was what he called teleportation, an instantaneous non physical movement of matter from one place to another. He saw this, if it existed, as a kind of vestige of the great forces that, in the formation of the planet, had shifted islands, mountains, and whole continents. Teleportation, he suggested, might account for the sudden appearance of objects in the séance room, mysterious disappearances of people and things, and falls of creatures from the sky.

Satirical attack

Fort's second book was *New Lands* (1923). This was partly a satirical attack on what Fort saw as the arrogance of astronomers. In view of the inevitable

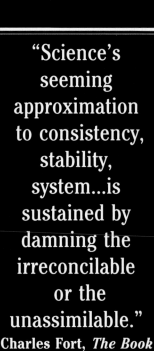

"Science's seeming approximation to consistency, stability, system...is sustained by damning the irreconcilable or the unassimilable."
Charles Fort, *The Book of the Damned*

> ## "I cannot say that truth is stranger than fiction, because I have never had acquaintance with either."
>
> **Charles Fort,**
> *Wild Talents*

absence of any empirical evidence for their claims, he found their categorical assertions about the exact location and distance from the earth of astral bodies somewhat ridiculous — doubly so, taking into account their past mistakes. But, "so great is the hypnotic power of astronomic science," he wrote, "that it can outlive its 'mortal' blows by the simple process of forgetting them."

Alien spacecraft

New Lands also suggested that the strange lights and shapes that so many around the world claimed to have seen might be those of alien spacecraft — something again dismissed out of hand by astronomers. Thus, a quarter of a century before the term "flying saucer" was coined in 1947, Fort had begun to explore the field of UFO's.

New Lands was followed by *Lo!* (1931), which was mainly concerned with accounts of rains of strange objects — including coins,

"manna," frogs, worms, shells, and snails — and many other possible indications of the existence of teleportation.

Sea serpents and UFO's

Fort's last book was *Wild Talents* (1932), which was in the main an account of incredible coincidences, psychic powers, unaccountable disappearances, and manifestations of the occult.

Fort's achievements were remarkable. Single-handedly, he dragged into the arena of serious study subjects previously considered to be of no interest to an intelligent person. Besides the topics already mentioned, Fort included lake monsters and sea serpents, wolf children, and tribes of giants and dwarfs.

Fort was also one of the chief initiators of all modern widespread interest in UFO's and the idea that extraterrestrials are possibly visiting the planet earth. His work has also inspired many notable science-fiction writers, including Robert Heinlein and James Blish.

Chaos theory

Fort accurately anticipated a modern development in physics: what is known as chaos theory, which suggests that apparently insignificant, scarcely discernible activities in one part of the world can, in theory at least, initiate a train of events culminating in major events in another continent. Half a century earlier, in *Wild Talents*, Fort had written: "Not a bottle of catsup can fall from a tenement-house fire escape... without affecting...the demand, in China, for rhinoceros horns...."

Today there are various publications devoted to furthering Fortean investigation. In the United States, for example, there are the *Info* (International Fortean Organization) *Journal* and *Pursuit*, published by the Society for the Investigation of the Unexplained; and in Britain *Fortean Times*. Such publications are testimony to the continuing influence of a truly unusual and pioneering thinker.

RECLUSIVE RENEGADE

Pugnacious opponent of scientific arrogance, bold, original historian of the bizarre, Charles Fort was, in his private life, a mild-mannered recluse.

CHARLES HOY FORT WAS BORN in Albany, New York, in 1874. He did not get along with his autocratic father, and this may have been the cause of the rebellion against conventional beliefs that is so apparent in his books. As a boy he wanted to become a naturalist but instead became a journalist. After working on two newspapers, then traveling around the world, Fort returned to New York in 1896. There he lived in poverty for a number of years, supporting himself and his wife, Anna, through journalism and odd jobs.

Voracious reader

Fort maintained a lifelong practice of voracious reading and prolific note-taking. His reading consisted mainly of scientific publications and of newspapers from all over the world. His notes were made on numerous scraps of paper, which he filed in his room, in pigeonholes on the wall or in cardboard boxes. By the age of 23, Fort had collected an astounding 25,000 notes on science's pretensions to infallibility, but as he put it, "they were not what I wanted," and he destroyed them.

Fort soon resumed his extensive research, and over the years accumulated tens of thousands of notes on bizarre happenings all over the world. In 1916, at the age of 42, he received a legacy from an uncle, which freed him from the worries of struggling for a living. He then began writing *The Book of the Damned*, based on his research.

Wealth of material

Fort's friend, the novelist Theodore Dreiser (1871–1945), was fascinated by the manuscript and in 1919 persuaded his publishers, Boni and Liveright, to take on the book. Despite critical acclaim, however, and Dreiser's support, Fort fell into a depression the following year and once more destroyed all his notes. Fort then sailed for London, where, apart from a brief spell back in New York, he and his wife lived until 1929. The library of the British Museum provided Fort with a wealth of new material on the unexplained, some of which he used to write *New Lands* (1923).

> **"How do geologists determine the age of rocks? By the fossils in them. And how do they determine the age of the fossils? By the rocks they're in."**
>
> **Charles Fort, *Lo!***

In contrast to his combative, imaginative writings, Fort was a quiet, kindly, retiring individual who spent most of his mornings and afternoons working and his evenings at the movies with his wife.

True legacy

Back in America, Fort published two more books: *Lo!* (1931) and *Wild Talents* (1932). In 1931, the year before Fort's death, the novelist Tiffany Thayer founded the Fortean Society, which continued meeting until 1959. In 1932 Fort became ill with an unspecified weakness and died in Royal Hospital, New York, on May 3 that year. He left behind tens of thousands of notes, which are now housed in the New York Public Library. But Fort's true legacy was the new approach to the bizarre and unexplained that he pioneered. The English novelist John Cowper Powys (1872–1963) wrote of him: "He creates that curious awe in the mind, in the presence of this inexplicable universe, which Goethe in *Faust* declares to be one of man's noblest attributes."

> **PUBLISHERS' NOTE**
> *Charles Fort made a point of citing the sources for the often incredible stories on which his books were based. In the same way, we provide, on pp. 140–141, a page-by-page breakdown of the source material used in the preparation of this volume.*

CHAPTER ONE

EXCEPTIONAL ENERGIES

People bursting into flames for no apparent reason, fish falling from the sky, flattened circles suddenly appearing in crop fields....Is there a perfectly rational explanation for these and other strange phenomena — or might they simply be the result of forces as yet unexplained by science?

On the morning of July 2, 1951, at 8 A.M., Mrs. Pansy Carpenter, the owner of an apartment house in St. Petersburg, Florida, took a telegram to the rooms of a tenant, Mrs. Mary Reeser, a 67-year-old widow. The landlady had last seen Mrs. Reeser late the previous evening sitting in an armchair in her kitchenette, wearing a nightgown, housecoat, and slippers, and smoking. To Mrs. Carpenter's alarm, she found the knob on the outer door of the

two-room apartment too hot to touch. The two painters she summoned for help were greeted by a blast of hot air when they opened the door; but, apart from a little smoke, the only sign of fire was a small flame on a partition dividing the bed-sitting-room from the kitchenette. Mrs. Reeser's bed had not been slept in.

Horrifying sight

When firemen were called, they came upon a horrifying sight in the kitchenette: on the floor was a blackened circle about four feet across, containing some armchair springs, a shrunken skull, a fragment of spine, a charred liver, a small mound of ashes, and an unburned left foot in a slipper. Oily soot covered the ceiling and the walls down to a level about four feet above the floor. Two candles on a dressing table and some plastic electrical wall outlets had melted, and a mirror had cracked. Yet nearby draperies, linens, and newspapers were untouched. The damage to the rest of the apartment was minimal.

Assistant Fire Chief S. O. Griffith and Coroner Edward T. Silk were bewildered by what they found. So, too, were the FBI and arson experts. The case appeared to be a modern example of a phenomenon reported regularly down the centuries: spontaneous human combustion, or unintentional self-ignition without the seeming involvement of any external agency or energy source. Modern medical science, and other traditional scientific investigatory agencies, have refused to recognize the existence of any such phenomenon. Yet, if the facts and

> # On the floor was a blackened circle about four feet across, containing some armchair springs, a shrunken skull, a fragment of spine, a charred liver, a small mound of ashes, and an unburned left foot in a slipper.

accounts are valid, no one has been able to offer any satisfactory explanation for reports of deaths like that of Mrs. Reeser.

The authorities called in Dr. Wilton M. Krogman, a physical anthropologist at the University of Pennsylvania and a leading expert on deaths by fire, to investigate the Reeser case. He quickly established that neither lightning nor an electrical fault nor arson was responsible for starting the fire. He also dismissed the possibility that Mrs. Reeser might have accidentally set herself alight while she was smoking; for one thing, according to Dr. Krogman, neither her clothing nor her chair was particularly flammable.

High temperatures

Dr. Krogman did conclude, however, that whatever form of conflagration had incinerated Mrs. Reeser, it must have reached a temperature of more than 3,000°F (1,650°C) because it is only at such heat that bone disintegrates. Yet this is a tremendously high temperature — greater even than that of a crematory, where bones have to be pulverized after incineration. What is more, crematory furnaces require copious amounts of fuel to maintain their high temperature. Thus

Indonesian incineration
A cremation pyre blazes on the island of Bali, Indonesia. To destroy a body completely and rapidly by fire, a tremendously high temperature is required.

it is very difficult to understand how the human body, with its high water content, might provide enough fuel to bring about such enormous heat. And how does the incendiary ring of a so-called spontaneous human combustion remain so clearly circumscribed, with the result that highly flammable materials located nearby emerge completely unscathed?

Dr. Krogman failed to provide any obvious solution to the mystery surrounding Mrs. Reeser's death. "I regard it as the most amazing thing I've ever seen," he declared. "As I review it, the short hairs on my neck bristle...."

***Mysterious death**
Officials inspect the kitchenette of the apartment in St. Petersburg, Florida, where Mrs. Mary Reeser (below) burned to death. Despite the high temperature of the fire, flammable objects around her remained virtually undamaged.*

Survivors

There are many recorded cases of people surviving apparent spontaneous human combustion. The French physician Richond des Brus, writing in *Annales de la Société d'Agriculture du Puy* in 1827, described a case of alleged burning that was confined to the hands. A young man who helped smother his brother's burning clothes set his own hands alight. They burned for hours with bright blue flames, which were finally extinguished by frequent immersion in water.

In another case, on November 12, 1974, traveling clothing salesman Jack Angel parked his mobile home at the Ramada Inn in Savannah, Georgia, and went to bed. When he woke up, four days later, he found, he said, his right hand burned black, a hole in his chest, and burns on his legs, groin, and back.

Internal burns

Angel experienced no pain, and the sheet on his bed was undamaged. In the hospital, doctors stated that his burns were internal — they had developed from the inside outward — but they could offer no explanation for this. The mobile home was inspected, but nothing that could explain or cause the burns was found.

On October 9, 1980, naval airwoman Jeanna Winchester was driving along Seaboard Avenue in Jacksonville, Florida, with Lesley Scott, one of her woman friends. Suddenly, for no discernible reason, Winchester is reported to have burst into flames. She began to scream: "Get me out of here!" Her passenger tried frantically to beat out the flames with her hands, but the automobile ran out of control and into a telegraph pole. Before the fire died out, it had burned more than 20 percent of Winchester's body surface — but she survived. Patrolman T. G. Hendrix, who arrived at the scene of the accident, said: "There was no other fire damage."

Burst into flames

On the night of May 25, 1985, as he walked along a road in Stepney Green, London, Paul Hayes, a young computer operator, suddenly became enveloped in flames from the waist up. He covered his eyes, fell to the sidewalk, and curled up in a ball. Thirty seconds later his ordeal was over. Somehow he managed to stumble to the nearby London Hospital, where he received treatment for burns to his head, neck, arms, and hands.

None of the people involved in these astonishing modern cases was smoking a cigarette at the time, nor did there appear to be any other apparent cause for their suddenly bursting into flame.

BURNING QUESTIONS

Are the fires that allegedly consume a human body, while causing no damage to the immediate surroundings, the result of spontaneous human combustion? Or is there a rational explanation for them?

THE FIRST MEDICAL WRITER to seriously consider the idea of spontaneous human combustion was the 17th-century Danish anatomist Thomas Bartholin. In his book *Acta* (1673) Bartholin described the case of a Parisian woman reportedly found burned to ashes except for her head and fingertips. In another incident in the following century, a Frenchwoman, Nicole Millet, of Rheims, was found in her kitchen almost totally consumed by fire. Her remains were in an undamaged chair. Her husband was accused of murder, but the physician Claude-Nicolas Le Cat persuaded the court that the woman's death was due to spontaneous human combustion. This case inspired his compatriot Jonas Dupont to write *De Incendiis Corporis Humani Spontaneis* (1763), a pioneering work on the phenomenon that drew together all known evidence.

Flammable spirits

A popular 19th-century theory to explain the mystery of such deaths was that the victims were drunkards whose bodies had become saturated over the years with flammable spirits. Nevertheless, this theory does not stand up: long before a victim reached the state of supersaturation necessary to blaze fiercely, he or she would have died of alcohol poisoning.

In modern times, Ivan T. Sanderson, in *Investigating the Unexplained* (1972), has expressed a belief in the validity of spontaneous human combustion and advanced a scientific explanation for it. Sanderson argues that what may be called a negative state of mind — brought about by illness or loneliness, for example — can adversely affect the body's metabolism.

Metabolic changes

In the December 1957 issue of the scientific journal *Applied Trophology*, cited by Sanderson, the theory is advanced that, as the result of one such metabolic change, phosphagens (compounds found in animal and human tissue that provide a reserve of chemical energy) might well build up in the muscle tissue of sedentary people. Such an accumulation, it has been suggested, might possibly make their bodies "subject to ignition," so that they might "burn like wet gunpowder in some circumstances." In addition, Sanderson argues that the combustion needed might be triggered by such phenomena as ball lightning or cosmic rays.

However, Livingston Gearhart, professor of music at the University of New York, Buffalo, has put forward

---

Nicole Millet, of Rheims, was found in her kitchen almost totally consumed by fire. Her remains were in an undamaged chair. Her husband was accused of murder....

another hypothesis. Writing in the paranormal magazine *Pursuit* in 1975, Gearhart linked many cases of claimed spontaneous human combustion with peaks in the earth's magnetic field. He suggested that such high magnetism might produce ball lightning, which in turn could generate radio short waves similar to those used in microwave ovens. And such speculation, though seemingly fanciful, is supported by the fact that some victims of mysterious incineration do seem to have been burned from the inside outward, as if by microwave radiation.

Another possible cause, and one sometimes advanced to explain the sudden ignition of a human body into flames, is its capacity to produce static electricity. Some people can reportedly generate as much as 30,000 volts.

Human candle

However, the more outlandish explanations for spontaneous human combustion are intriguing. A British television program *Q.E.D.*, on April 26, 1989, argued that bizarre incinerations of the human body can be explained quite simply, by the known laws of physics and chemistry. The speculative scenario goes something like this: The victim, having fallen unconscious or even died from natural causes, catches fire from some obvious heat source, such as a cigarette, fire, or stove. As the body burns, suggested Dr. Douglas Drysdale, Director of the Unit of Fire Safety Engineering at Edinburgh University, the clothing on it becomes saturated with melting body fat, and acts as a

Alcoholic ablaze
A drunkard apparently igniting spontaneously, as portrayed by the 19th-century English illustrator George Cruikshank.

wick, turning the unfortunate victim into a kind of grotesque human candle.

The program demonstrated that, in a well-insulated room, a burning body would, in about four minutes, lower the oxygen level in the atmosphere to 16 percent. This would turn the fire on the body to a smoldering one. With such a fire, bones could be reduced to ash by a temperature of 932°F (500°C), given a long enough period — 12 hours or so. (Drysdale pointed out that the reason higher temperatures were needed in a crematory to destroy bone was that the process of cremation had to be completed in a short space of time.)

Heavy condensation

The program suggested that the initial heat of the fire vaporizes the body's water content. This might explain why heavy condensation is found on the walls of rooms where deaths from spontaneous combustion are supposed to have occurred.

The heat of the smoldering fire, so the speculation continues, would not be great enough to burn nearby linen or newspapers, but hot gas would be of a sufficiently high temperature to melt plastic articles and heat up metal objects such as door knobs. This theory may provide a speculative answer for many of the more puzzling aspects of so-called spontaneous human combustion. But it does not offer a satisfactory explanation for the cases of people who, it appears, have burst into flames while not close to any source of flame or ignition, and who have happily survived to tell the tale.

> As the body burns...the clothing on it becomes saturated with melting body fat, and acts as a wick, turning the unfortunate victim into a kind of grotesque human candle.

THE POWER TO IGNITE

Some people, most often children, seem to be able to set objects around them on fire, without using normal means of ignition.

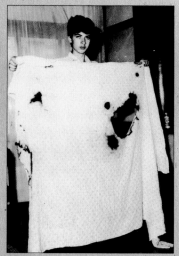

Fire risk
Since Benedetto Supino, an Italian boy, was nine years old, objects around him have apparently ignited spontaneously. Here, age 16, he holds a sheet in which, it is claimed, holes were mysteriously burned while he was asleep.

BENEDETTO SUPINO, nine-year-old son of a carpenter in Formia, Italy, was reading a comic book in a dentist's waiting room in 1982, when the comic caught fire. On another day he awoke to find his bedclothes on fire, and in another reported incident he stared at a plastic object held by his uncle until it burst into flame. Furniture and fittings smoldered in his presence, and electrical equipment functioned erratically. Leading physicians, including Dr. Giovanni Ballesio, dean of physical medicine at Rome University, examined the boy but could discover nothing medically unusual about him.

Another reputed "fire boy" in recent times was the 13-year-old Russian schoolboy Sascha K., from Yenakiyevo in the Ukraine. Beginning in November 1986, he was reported to have unwittingly started more than 100 domestic fires, setting alight clothes, furniture, carpets, and electrical equipment, and causing light bulbs to explode. Following these incidents, Sascha K. was taken to a hospital for observation, and there the clothing of a boy who shared his room suddenly caught fire.

Evil eye?

One famous case of apparently unwitting fire-starting was that of a Scottish nursemaid, Carole Compton. On December 12, 1983, Compton, then age 20, was tried in Leghorn, Italy, for arson and the attempted murder, the previous year, of Agnese Cecchini, the three-year-old girl then in her charge. Compton was accused of setting fire to two mattresses next to the girl's crib. There had been a similar fire the previous night in the bedroom of Agnese's grandfather. On both occasions Compton had been dining with the family, which led her ward's grandmother to claim that the nursemaid had started the fires by using the "evil eye." A previous employer of Compton testified that there had been unexplained fires in her household, that her two-year-old son used to cry out that his nursemaid was burning him, and that her maid had complained that pictures of the Madonna kept falling off walls and vases tumbled off tables in Compton's presence.

Strange type of arson

Forensic experts at Compton's trial testified that the mattresses had burned only from the top downward, in a way they had never seen before. Professor Vipo Nicolo of Pisa

> Forensic experts at Compton's trial testified that the mattresses had burned only from the top downward, in a way they had never seen before.

Nursemaid on trial
Scottish nursemaid Carole Compton, another young person with the apparent power to ignite objects without normal means of combustion, arrives at court in Leghorn, Italy, in December 1983, to face trial for arson.

University reported: "In my 45 years I have never seen a fire like it." Despite such circumstantial evidence, and perhaps because of its unusual nature, Compton was found guilty. Since she had already spent 16 months awaiting trial, however, she was released immediately.

Fiery farm

In another case, on August 6, 1979, smoke was observed billowing out of an abandoned farmhouse belonging to the Lahore family in Séron, in the Hautes-Pyrénées, France. The fire was quickly extinguished, but two more fires soon broke out in the Lahore's modern farmhouse. During the following month 90 more mystery fires occurred in the new farmhouse. Twenty gendarmes camped near the farm, but despite their round-the-clock vigilance, no arsonists were discovered. In a single day 32 separate fires sprang up among items of linen, clothing, or furniture. First there would be a smell of smoke and the linen, clothing, or furniture would show a circular charred spot. Items marked in this manner would then burst into flames.

The prefect of police, an investigating judge, and the head of a local analytical laboratory failed to discover any cause for the fires. Despite a lack of evidence, two young men were convicted of arson, but, like Carole Compton, they were released after the trial.

Wings burned off

Sometimes whole localities can be plagued by fires of unknown origin. In the summer of 1983 both the small West Virginia coal town of Wharncliffe and nearby Beech Creek suffered in this way. The first clue that something was wrong was the fact that wasps in the area were walking, not flying: their wings had been burned off. On May 27 a house burned down, and on June 9 another house was destroyed in the same way. Four days later, eight separate fires broke out in four hours in the house of a lay minister, Eugene Clemons; flames shot six inches out of electrical sockets, and as a precaution, the local power company cut all lines into the house and the church next door. The next day there were more fires in the church basement.

Microwave radiation?

While Clemons's wife was driving undamaged clothing from their house to that of her mother, Chloe Kennedy, in Beech Creek, some of it burst into flames while in the trunk of the car. However, the clothing showed no signs of any flammable agent, and the burn patterns were unlike any known to the experts who investigated the case. The West Virginia fire marshal's office was convinced that the fires were arson, but it failed to find a culprit. The locals' theory was that the fires were caused by microwave radiation from the nearby Norfolk and Western Railroad communications tower on Horsepen Mountain. But state health department tests failed to detect the presence of any such radiation.

First there would be a smell of smoke and the linen, clothing, or furniture would show a circular charred spot. Items marked in this manner would then burst into flames.

INTO THIN AIR

When someone disappears, there is often a straightforward explanation. Yet in some cases the vanishing appears to be so bizarre that it defies all logic. Is it possible that there are forces at work in nature that we cannot explain?

*I*T WAS SOME TIME AFTER MIDNIGHT on February 26, 1985, when Richard Brownell and his fiancée, Sandra O'Grady, left a bar in Newport Beach, California, with a man who, they told friends, was flying them to Las Vegas to visit the casinos. An hour later a single-engine Cessna 152 two-seater plane plunged into the Pacific off Newport Beach. The bodies of Brownell and O'Grady, neither of whom could fly a plane, were found strapped to the seats, but there was no trace of the pilot. No engine faults were discovered, and three intense ocean searches failed to find another body.

There was no trace of the pilot. No engine faults were discovered, and three intense ocean searches failed to find another body.

Investigators looking for the third person reported that a car belonging to the plane's owner was discovered near to the plane's tie-down spot at John Wayne airport, but the owner could not be found. Nor could investigators establish that the owner of the plane was in fact the pilot who went off with the two victims.

Mystery pilot

The police believed that there may have been an airborne mugging: the deceased had had about $3,000 on them when they had set out, but there was only small change on the bodies. Although the plane was only a two-seater, there was a cramped space behind the seats where another person, if he had been small, could have hidden. If this did happen, how could the third passenger have escaped? The plane did not carry parachutes, and in any case there would not have been room behind the seats for one if someone had been hiding there. "It's getting to be a mystery," said Lt. R. Olson of the Orange County Sheriff's Department.

Missing motorist

Equally mysterious was the disappearance of Graham Marsden, an unmarried businessman of Poole, Dorset, in England. At 5 A.M. one day in January 1989, he filled up his red Volkswagen Polo at the Rownhams Service Station on the M27 near Southampton. After paying his

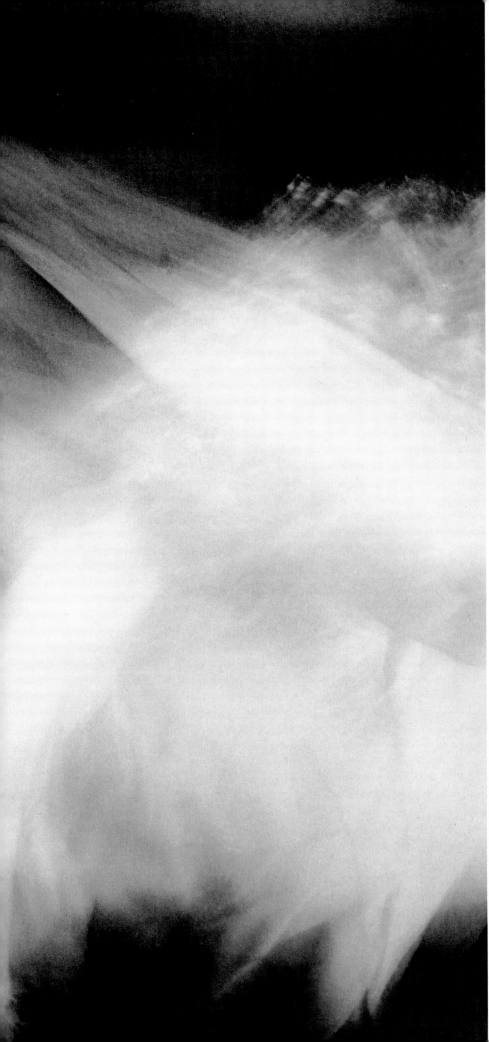

bill, he asked the way to the men's room, and was last seen walking toward it. After an hour, the gas station attendant went to look for the man. Finding no trace of him, he called the police, who searched the surrounding woods with tracker dogs. However, there was no sign of Marsden.

The police found that Marsden had no personal or financial worries. "If he planned to disappear," mused a police spokesman, "why did he fill his car with petrol first and leave it at the pumps? And why choose a service station miles from anywhere? We are just mystified."

> "If he planned to disappear," mused a police spokesman, "why did he fill his car with petrol first and leave it at the pumps? We are just mystified."

On July 14, 1990, an excursion organized by the Orkney Heritage Society and the Royal Society for the Protection of Birds landed 88 ferry passengers on Eynhallow, an uninhabited islet off the coast of Scotland. Only 86 returned. Police and coastguard teams searched the tiny island of Eynhallow extensively, and a rescue helicopter with heat-seeking equipment scanned the area. But no trace of the missing persons was found.

The vanishing isle
The Scottish Isles have long been renowned for their mystic and occult connections, and Eynhallow is steeped in sinister legend: it was once known as the "vanishing isle." It is reported, so the story goes, that no mice, rats, or cats can survive there, and blood is said to flow from any corn cut after sunset. Yet such fanciful traditional myths probably had nothing to do with the vanishing ferry passengers. But when all logical explanations are exhausted, the human mind tends to consider other alternatives, including the paranormal. How else, we struggle to ask, could two people apparently vanish into thin air?

THE VANISHING SHIP

Could science ever succeed in making matter invisible? According to one self-professed witness, the U.S. Navy accomplished this astonishing feat in 1943 when it allegedly made a destroyer vanish into thin air.

"THE RESULT OF THE EXPERIMENT was complete invisibility of a ship, destroyer type, *and all* its crew, while at sea (Oct. 1943). The field was effective in an oblate spheroidal shape, extending one hundred yards...*out* from each beam of the ship. Any person within that sphere became vague in form....Somehow, also, the experimental ship disappeared from its Philadelphia dock and only a very few minutes later appeared at its other dock in the Norfolk, Newport News, Portsmouth area...the ship then *again* disappeared and went *back* to its Philadelphia dock."

Einstein's theory

The above is an extract from a letter written on January 13, 1956, by Carlos Miguel Allende (also known as Carl Allen), a former seaman, to Dr. Morris Jessup, an astronomer and author of *The Case for the UFO* (1955). The experiment, claimed Allende, was an application, by a scientist called Dr. Franklin Reno, of Einstein's unified field theory (a theory that attempts to connect the fundamental forces of nature). Reno, Allende claimed, had successfully connected the field of gravity with that of electromagnetism.

Disappearing destroyer

Allende stated that he witnessed the disappearance of the destroyer, the U.S.S. *Eldridge*, while on board a merchant ship, the S.S. *Andrew Furuseth*. However, the appearance of the *Eldridge* in the Norfolk area was something that Allende claimed he had only read about, in a Philadelphia newspaper.

According to Allende, the success of the experiment was marred by terrible side effects on the *Eldridge* crew. Once, during a dockside bar brawl, some of these crewmen, contended Allende, vanished into thin air. He further claimed that there was a report about

this incident which had appeared in a Philadelphia newspaper, sometime in 1944–46. In the best-selling book they wrote about this mystery, *The Philadelphia Experiment* (1979), Charles Berlitz and William Moore reproduce a newspaper clipping of such an account. Yet the clipping does not bear any newspaper name or date, and its column width is greater than that of any Philadelphia newspaper published in the 1940's.

Naval records

Dr. Jessup seems to have believed Allende's story and conducted research on it. Naval records show that Allende was, as he had claimed to be, aboard the *Andrew Furuseth* in October 1943. But these records show that the only possible date on which that ship and the *Eldridge* could have been (though there is no evidence that they were) in the same vicinity was August 16, 1943. In addition, no other former crew members of the *Andrew Furuseth* have at any time corroborated Allende's story.

Alleged mastermind

There is also no mention of Dr. Franklin Reno, the alleged mastermind behind the experiment, in any reference works. Berlitz and Moore claim that his name was a pseudonym, that they traced and interviewed him, and that he died several months later. The authors say "Reno" told them he was a top scientist for the U.S. military, involved in research into the use of electromagnetic fields to deflect torpedoes and mines.

The U.S. Navy has always denied that the experiment took place, and there is no other evidence that this or similar experiments occurred. It seems probable that the Philadelphia Experiment existed only in the mind of Carlos Allende.

Screen version
Berlitz and Moore's book The Philadelphia Experiment *(1979) inspired the 1984 science fiction film of the same name.*

THE MAGNETIC CLOUD

At the beginning of the century, Philadelphia was at the center of another curious maritime episode. On August 1, 1904, while the British steamship *Mohican* was in port, a story appeared in the *Philadelphia Inquirer* that the previous month, while making for the Delaware Breakwater, the vessel had been enveloped in a strange magnetic cloud.

Immovable chains

According to Captain Urquhart, the master of the *Mohican*, the cloud — gray, with glowing spots — was so dense that nothing could be seen beyond the decks and everything seemed to be a mass of glowing fire. The compass spun wildly, and iron chains on deck were magnetized and could not be moved. "The hair on our heads and in our beards stuck out like bristles on a pig..." Urquhart was reported as saying. "All the joints of the body seemed to stiffen...there was a great silence over everything that only added to the terror."

After half an hour, it was claimed, the cloud lifted and moved off over the sea. No explanation was ever offered for this alleged bizarre happening.

OUT OF THE BLUE

The sudden appearance of objects in unexpected places is one of history's most baffling and bizarre phenomena.

FALLS OF AMPHIBIANS, FISH, AND ARTIFACTS from the sky have been recorded from earliest times. However, since the ancients could no more explain rain, hail, and snow than they could account for precipitations of fish or stones, they regarded such manifestations as being of divine origin. With the dawning of modern scientific thought in the 17th century, any falls of objects from the sky were neatly divided into either the meteorological kind, which could now be understood, or the remarkable kind, which were now dismissed as illusions or hoaxes.

Scientific blunder
The latter category included meteorites. In 1772, a committee of the Académie Française investigated the fall of a stone that had landed with a loud explosion in Luce, a small village in the province of Maine. The committee said that it was a self-evident absurdity that stones could exist in the sky. As a consequence, they concluded that the one in Luce could not have fallen

Cardan put forward a new theory: that a whirlwind or waterspout lifted up the fish or frogs and deposited them at some distance away.

but must have been unearthed by lightning, which must have caused the explosion and left the stone hot. This attitude toward meteorites was abandoned in the 19th century, in the face of the overwhelming evidence for their existence. A telling lesson perhaps for those who discount the existence of paranormal phenomena without even examining the evidence.

Fertilizing frogs and fish?
The most common falls of creatures from the sky have been of frogs and fish. The earliest record of such falls comes from the Roman naturalist Pliny the Elder (A.D. 23–79), who believed that they were due to frog and fish "seed" in the soil being brought to life by rain. When this idea was discounted, it was suggested that rain flushed out frogs from their hiding places in the ground. Then in the 16th century the Italian physician, mathematician, and astrologer Jerome Cardan (1501–76) put forward a new theory: that a whirlwind or waterspout lifted up the fish or frogs and deposited them at some distance away.

Amphibious landings

Frog falls are one of the most commonly reported of weird precipitations. The May 1958 issue of Fate *magazine contained one such true-life story.*

clearly defined areas. Neither of these circumstances fits in with the common conception of a whirlwind.

"A Catalogue of Meteorites and Fireballs, from A.D. 2 to A.D. 1860," first published in 1860 by the English astronomer R. P. Greg, listed many bizarre items said to have fallen in the company of meteorites, including blood and "gelatinous substances." But from about 1865 scientific literature gradually excluded reports of such falls from the sky and of other paranormal phenomena, damning them as old wives' tales.

Eels, ants, worms

Reported precipitations of animals and objects returned to the realm of public record in the early 20th century, thanks to the work of Charles Fort. In *The Book of the Damned* (1919) he bombarded the reader with accounts not only of rains of frogs and fish but also of falling ice, ashes, mud, sulfur, bricks, cinders, stone axes, nails, iron chains, snakes, eels, ants, worms, meat, seeds, nuts, and much else. And in modern times well-documented accounts of such precipitations have continued to accumulate. The following is a small sampling of the most dramatic.

Frog falls

During a thunderstorm in the summer of 1926, Mr. W. A. Walker, a golf caddy of Evansville, Indiana, watched in absolute amazement, with the golfers on the

This hypothesis has been attacked on the grounds that a whirlwind or waterspout, not being selective, would also gather up mud, stones, plants, and other matter and deposit these too. And when frogs or fish have been observed to fall, nothing else has been reported to have come down with them. Another objection has been that some falls have taken place on clear, still days, and others have been deposited on small,

GONE WITH THE WIND

On January 3, 1978, several skeins of pink-footed geese were caught by a tornado moving across the county of Norfolk in England and were swept up to a great height. Over the next hour 136 of the geese dropped to earth dead, 105 of them falling into fields near Wicken Farm, Castleacre. Pink-footed geese are normally found over the Wash, 15 miles or so away.

Powerful whirlwind

Is it possible that other reported strange falls of objects from the sky have been caused by tornadoes? On the tornado intensity scale produced by TORRO (Tornado and Storm Research Organization) of Oxford, England, even T6 ("moderately devastating") tornadoes, whose wind speeds measure up to 186 m.p.h., are capable of lifting heavy motor vehicles and roofs into the air. T10 ("super") tornadoes, whose winds can be in excess of 280 m.p.h., can rip buildings from their foundations, lift them up in the sky, and deposit them a considerable distance away.

A tornado is the most powerful type of whirlwind. It is formed when a powerful current of warm air rises to meet a fast-growing mass of cloud. The updraft begins rotating to form a funnel of water droplets. This funnel cloud descends from the base of the main cloud, to which it appears to be attached like a tail. Where the foot of the funnel remains above ground, it can pass right over buildings and people, leaving them unharmed. If it reaches the ground it becomes a tornado; if it reaches water, it turns into a waterspout.

Hidden in the cloud mass

While it is merely a funnel cloud, a latent tornado remains hidden in the main cloud mass. Moreover, the tornado or waterspout itself usually affects no more than a few square miles and typically its effects lasts only minutes, or even seconds. These two facts may offer some explanation as to why, if tornadoes or waterspouts are in fact responsible for these bizarre precipitations, they might pass quite unnoticed.

Prodigious fall
An engraving of a fish fall from the 16th-century Book of Prodigies.

course, as "thousands of frogs came right down with the rain...about the size of nickels, they were alive and jumping."

In October 1987, during a torrential rain storm, a host of minute rose-colored albino European common frogs (*Rana temporaria*) fell to earth around Stroud,

The frogs bounced off umbrellas and pavements and hopped off to nearby streams and gardens.

in Gloucestershire, England. Two further falls had been witnessed in the nearby town of Cirencester only a fortnight earlier, at about the same time that a fine sand, allegedly airborne from the Sahara Desert, was deposited all over the country. A fourth fall of frogs also took place in the Charlton Kings area of Cheltenham, in the same county. According to one witness, the frogs

bounced off umbrellas and pavements and hopped off to nearby streams and gardens. The Gloucester Trust for Nature Conservancy investigated the theory that these creatures had arrived with the sand, possibly "carried in atmospheric globules of water across land and sea."

Pink precipitation
This pink albino European common frog is one of the many that fell in 1987 on the county of Gloucestershire, England.

BOMBS GLACÉS
Falls of ice have been recorded throughout history. One massive block, 15 feet long, 6 feet wide, and 11 feet thick, is said to have fallen in France during the reign of the emperor Charlemagne (*c.* A.D. 742–814). Another block of ice, with a circumference of about 20 feet and weighing about half a ton, is reported to have fallen in 1849 on Ord, a village on the island of Skye, Scotland.

Crashed through roof
On April 25, 1989, a volleyball-sized chunk of ice crashed through the roof of a house in Portland, Oregon. It was milky white and smelled slightly of sulfur. In August of the same year a 375-pound block of blue-green ice descended from a cloudless sky onto a field near Belgrade, Serbia. In April 1990 a lump of ice the size of a football also fell, from a clear sky, into a garden near Lake Vättern, Sweden. It was white with gray, brown, and lilac markings.

Plane truth
Such present-day ice falls are popularly believed to originate from aircraft with either faulty de-icing equipment or leaking lavatories. But atmospheric physicist James McDonald, writing in *Weatherwise* magazine in June 1960, suggested that, of 30 selected ice falls in the 1950's, only two could be attributed to planes.

On the evening of August 20, 1984, a vast number of freshwater shad fell from a cloudless sky around Mr. and Mrs. A. D. Ellmers and their neighbor Walter Davies, as they stood in the driveway of the Ellmers' house in Bonita, California.

Shad shower

The startled witnesses described the fish as landing with "harsh wet splats." Shad were known to inhabit the Sweetwater reservoir, about two miles away, but no waterspout capable of lifting the fish had been reported on that particular evening.

The following year, on April 21, during a heavy rainstorm in the area of St. Cloud, Minnesota, more than 100 miles inland, two residents were perplexed to see "a white thing with five legs" in their backyard. It was a starfish. When they turned on the outside lights, they saw more on their lawn. Others were found in a neighboring garden.

> **The Haythornwhites were awakened in the night by a tattoo on the roof that lasted for an hour or more. The next morning they found at least 300 out of season and quite fresh apples on their back lawn.**

The starfish were identified as a species that usually inhabited the ocean off Florida, more than 1,000 miles away.

Thousands of dead sardines fell "like a sheet of silver rain" on February 6, 1989, around the house of Harold and Debra Degen of Rosewood, in Queensland, Australia. The Degens gathered a bowlful for their cat and a few more to keep as souvenirs; the rest were "gobbled up by kookaburras" (a type of Australian bird).

Heavenly windfall

On an otherwise peaceful night in November 1984, Derek and Adrienne Haythornwhite, of Accrington, Lancashire, England, were awakened in the night by a tattoo on the roof that lasted for an hour or more. The next morning they found at least 300 out of season and quite fresh apples on their back lawn, on the path, and in hedges. More were found in nearby gardens.

Raining cats and dogs
George Cruikshank, the 19th-century English artist, produced this illustration, entitled "Raining Cats, Dogs, and Pitchforks," as a satire on contemporary reports of bizarre precipitations.

APPORTS

Most materializations of objects at séances have proved to be fraudulent. Yet a few mediums do seem to be able to pluck unusual objects from the unknown.

Stanford apport: tortoise shell

IN THE ARCHIVES READING ROOM of Stanford University in California is a collection of apports, objects that have apparently been transported paranormally by a medium from a distant place to the séance room. The Stanford apports, which include shark's teeth, pottery fragments, a Roman lamp, a message on a slate, tortoise shells, and a human shoulder blade, were produced during the early part of the century by the once-celebrated Australian medium Charles Bailey (*c*.1870–1947) at various séances in his own country and at millionaire Thomas W. Stanford's mansion on Nob Hill in San Francisco. Stanford, a supporter of spiritualism and patron of Bailey, was the brother of the founder of Stanford University.

Placed in cage

Bailey produced his apports under what appeared to be controlled conditions. Members of the audience body-searched him before the séance, he was tied up in a sack with openings only for his hands, and then placed in a cage or in mosquito netting. The séance room was then plunged into darkness, and when the lights were switched on, one or more objects had mysteriously appeared within the cage or netting.

Stanford apport: human shoulder blade

Over the years the objects thus produced by Bailey were remarkably varied and unusual. They included, in addition to those items already mentioned, a live turtle, crab, snake, and fish, live birds, together with their nests, ancient coins, seaweed, and a leopard skin.

Big boots

Bailey's séances were sometimes attended by psychical investigators, and there were various indications that his apports were fraudulent. On one occasion, for example, a bird dealer reported having sold to Bailey birds of the same type that the medium soon afterward produced at a séance. At another session, when the light was switched on unexpectedly early, Bailey was discovered with one boot off. This, and the fact that the medium was a former bootmaker and wore boots with very deep heels, led to the suspicion that he concealed articles in them. In addition, it was believed by some investigators that Bailey swallowed and regurgitated various objects or secreted them in his intestinal tract.

In modern times, alleged apports — like claims of ectoplasm production (the fleshy substance that spiritualists claim appears as spirits take form), levitation, and other physical phenomena common to the séance room — have become rare. This may be due to the fact that reports of paranormal phenomena do appear to be subject to the vagaries of fashion. Another reason may be that psychical investigators now use high-tech equipment, such as infrared cameras, which can reveal activity taking place in the dark. This may have put an end to much of the deception once perpetrated by mediums.

Baskets of candies

In India, however, the swami (Hindu religious teacher) Sathya Sai Baba, a legend in his own country, has produced apports for more than 50 years without any evidence of fraud being detected. The objects that Sai Baba seems to pluck from thin air have included fruit, baskets of candies, foods, rings, prayer beads, and sometimes even objects on request, such as a statuette of a particular deity.

Dr. S. Bhagawantam, an Indian nuclear physicist and former director of the All India Institute of Science in Bangalore, has observed Sai Baba closely and is convinced his apports are genuine. Celebrated American magician Douglas Henning, however, and many others, both magicians and scientific investigators, believe that most of these paranormal phenomena could be achieved by clever legerdemain.

Stanford apport: Roman lamp

SEALED MOVES

A psychical society in Missouri claims that objects placed in sealed containers have frequently changed position, without what appears to be any detectable external influence.

Two seamless leather rings in the container become interlinked....

And a mixture of differently colored dyed peas sorts itself into single-color groups.

A LARGE FLORIDA COCKROACH in a sealed container mysteriously leaves it and turns up inside a light bulb. Two seamless leather rings in the container become interlinked. And a mixture of differently colored dyed peas sorts itself into single-color groups, as does a pack of cards, into suits. These and many other alleged examples of PK (psychokinesis — the movement of objects by psychic means) are claimed by members of the Society for Research into Rapport and Telekinesis (SORRAT) in Missouri. (Telekinesis is another term for teleportation, the paranormal movement of objects.) To substantiate its claims, the society has made widely available photographs and films of many such apparently paranormal happenings.

Previous experiments

Such results (achieved under controlled conditions) are by no means unique. In 1943, for example, the pioneering parapsychologist Dr. J. B. Rhine claimed that the results of experiments on the power of mental concentration over dice scores, conducted in his laboratory at Duke University, in North Carolina, were indisputable evidence of the reality of PK. Another notable example was the report in the 1960's by the Institute of Brain Research, in Leningrad, that the Russian psychic Nina Kulagina could apparently move objects without physical contact.

Tapping sounds

SORRAT was founded in 1961 by Dr. John Neihardt, an eminent professor of English literature at the University of Missouri, at his home, Skyrim Farm, near Columbia. The group he formed, originally 30-strong, carried out impressive experiments involving not only PK but also spontaneous writing, rapping (coded tapping sounds created by some allegedly unknown agency wishing to communicate messages), and other psychic activities. To induce PK, the group did not strain mentally, as, for example, Kulagina did, but sat around in a relaxed state waiting for the phenomenon to take place.

Together with W. E. Cox, J. B. Rhine's chief PK investigator, Neihardt built containers in which objects intended for PK experiments were placed. The sealed containers were made of glass, or glass-topped, to enable any movement of the objects to be filmed or photographed.

Moving dice

In some of the sealed boxes a pair of dice was placed on a layer of dried coffee grounds covering the base of the container. While the group was absent, one die would apparently move of its own accord: when the container was viewed again, the die, leaving a trail in the coffee grounds, would be found in a different position. In other boxes, containing a stylus and carbon paper, writing would be found on the white floor of the container, where the stylus had, it appeared, pressed on the paper.

Cox's box

Four years after John Neihardt's death in 1973, W. E. Cox constructed the first of several transparent mini-labs for SORRAT's experiments. The boxes were securely fastened, to make them more fraud-proof than the previous containers. The first mini-lab was set up at the home of SORRAT member, Dr. J. T. Richards, in Rolla, Missouri, where many of the experiments have since taken place.

Automatic filming

To record any movement of an object in the mini-lab during the absence of group members, Cox invented an automatic filming device. If any article moved, this activated a microswitch that turned on lights and a video camera, which shot

Test case
SORRAT experimenter W. E. Cox examines one of his securely fastened mini-labs, in which objects are claimed to have behaved in a paranormal way.

30-second sequences of film. Since 1979 all alleged PK phenomena at Rolla have been captured on film in this way. As well as insects, rings, peas, and cards, other unexpected objects said to have passed into or out of the mini-labs have included pipecleaners, toys, jewelry, pens, and matchbooks.

In an edition of the British television program *Arthur C. Clarke's World of Strange Powers,* screened on July 3, 1985, a balloon was shown being inflated without visible means inside Cox's mini-

> **In other boxes, containing a stylus and carbon paper, writing would be found on the white floor of the container, where the stylus had, it appeared, pressed on the paper.**

lab. In addition, the director sealed inside the container a letter, addressed to Arthur C. Clarke in Sri Lanka, together with some quarters. Two weeks later the letter, reportedly stamped with the correct postage and containing one of the quarters, is said to have reached Clarke.

Rapping at Rolla

Much of the reported PK phenomena at both Neihardt's and Richards's homes has been alleged by SORRAT to be the work of the same force (referred to by the group simply as "the agency") that is responsible for the rapping often heard by SORRAT members. W. E. Cox claims to have traced the source of the rapping — which is said to spell out messages in code — to the garden of Skyrim Farm. James Randi, the noted investigator of the paranormal, is of the opinion that the rapping is created by group members tapping their feet on the floor. But J. T. Richards claims that video film disproves this theory. According to SORRAT members, "the agency" has successfully rapped out, in a third of its experiments, the correct order of every ESP card (circle, plus sign, wavy lines, square, and star) in a sealed and randomly ordered deck.

Psychic poem

At Skyrim Farm, on Easter Sunday, 1991, SORRAT members claim that a poem was written by means of PK. A plastic box with a white cardboard floor, containing a pencil stub, was placed in Neihardt's study, and the group sat down to wait. After an hour or so had passed the telephone rang. It was a SORRAT member, Maria Hanna, living in Barstow, California, who said that rapping heard at her home indicated that there was now a poem in the box.

The group said that it did find a poem, written in pencil on the card, that was immediately recognizable as being in the style of the late Dr. Neihardt.

As well as the Arthur C. Clarke television film, an American film documentary by Alan Neumann, *The Psychic Connection*, has also shown apparently paranormal happenings taking place at Rolla.

Psychic writing?
Inside a SORRAT sealed container, a pen allegedly levitates and begins to spell out a message.

WORMHOLES IN SPACETIME

Is it possible that inexplicable materializations and disappearances of people and objects might be due to occasional discontinuities in the spacetime continuum – the three dimensions of space plus the fourth of time? This may in fact be the case, but the debate, like so many paranormal issues, remains highly theoretical and speculative.

Albert Einstein's General Theory of Relativity, the basis of modern physics, allows, to some degree, for the possibility that matter in one place or time might reappear in another. According to Einstein, the spacetime continuum is curved by the gravitational force of a great mass such as that of a star. In the years since Einstein's theorizing, some physicists, such as the Australian Roy P. Kerr, have suggested that the bending of spacetime might extend further – and lead to the creation in it of faults, or wormholes, which might connect separate, or perhaps distant, places and times in the universe. And so an object entering a wormhole might emerge somewhere else, either in the present, in the past, or in the future.

Black holes

The question of how and where wormholes might come into being is currently being studied. It is speculated that they might form in the centers of black holes, regions of incredibly high density with an intense gravitational field which exist at the centers of many galaxies and in the debris of exploded stars. Black holes swallow all matter or radiation that comes within a certain distance of them, cutting it

off from our universe forever. A wormhole (if such things exist at all) might lead out of the black hole to another part of our universe, or perhaps even to a completely different parallel universe.

Foamlike mass

Quantum physics, however, suggests that wormholes in spacetime need not be confined to astronomical distances and masses but may also exist on the subatomic level – in the world of protons, neutrons, and electrons – as something like a foamlike mass of holes that are continually appearing and disappearing. Theoretically, at least, these wormholes could also exist on some intermediate level that might affect everyday reality.

If so, might they be responsible for these appearances and disappearances that seem to flout the physical laws normally controlling the world around us? No one really knows.

Origin of a wormhole?
The center of NGC 4151, a spiral galaxy, shown here in an X-ray photograph, is believed by astrophysicists to be a black hole. It has been suggested that a wormhole in spacetime might form in such a black hole.

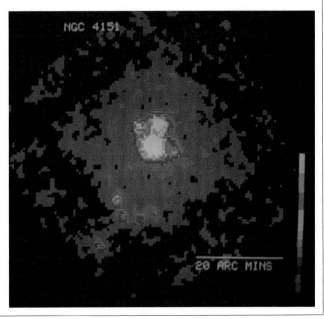

THE CROP CIRCLE ENIGMA

One of the strangest phenomena reported in recent times is the sudden formation of flattened areas, usually circular, in crop fields. The debate about how this happens, and the type of energy involved, continues among scientists and paranormal researchers.

*I*N ITHICA, NEW YORK, on October 20, 1967, a woman, known only as Mrs. Anna, saw a large glowing light in the night sky above some fields that she was driving by. The next morning she returned to the spot and found a crop circle 30 feet in diameter in a small swamp where "cattails had been squashed down and found to lie in a clockwise spiral pattern."

At dusk one evening in July 1988, in Gloucestershire, in the southwest of England, Tom Gwinnett had a similar experience. He said he saw what seemed like flashing electrical sparks zigzagging across the tops of a field of wheat and merging to form a large orange-yellow ball of light that spun above the crops. Returning the next day, he found a crop circle at the spot he had seen the ball of light.

> **Tom Gwinnett said he saw flashing electrical sparks zigzagging across a field of wheat and merging to form a large orange-yellow ball of light that spun above the crops.**

On the other side of the world, in the early hours of September 1, 1991, a circular depression was reported to have formed within the double-fenced grass compound of the Nippon Radio station in Tokyo, Japan. Considerable radio interference had also been reported during the period when it might have been formed.

There are hundreds of reliable reports annually of such circles. They have appeared all around the world, from the United States and Canada to the countries of Europe, as well as Russia, Japan, Australia, and New Zealand.

Sacred circles

Crop circles are not exclusively a 20th-century phenomenon. Mysteriously formed circles in crops have been observed since prehistoric times. Leading British crop circle investigator and meteorologist, Prof. Terence Meaden, sees crop circles as offering an

44

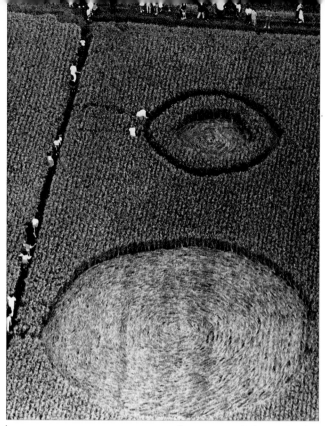

High-school pranksters
These circles appeared in a rice paddy in September 1990 in Fukuoka Prefecture, Japan. A few days later, a group of high-school students contacted the local police and admitted they had created these hoax circles.

explanation for the thousands of Neolithic and Early Bronze Age round burial mounds and stone circles found in Britain. Meaden argues that these were built on top of the sites of crop circles that may have been held as sacred.

Mowing devil

Crop circles have always amazed those who have witnessed them. In August 1678 in Hertfordshire, in the south of England, a four-page pamphlet was published that reported a mowing devil that appeared the night after a dispute between a mower and a farmer. In the heated argument that ensued, the farmer declared that he would rather "the Devil himself should Mow his Oats" than the poor mower. He was taken aback when a strange light was reportedly seen in the disputed field that night and a circle was found there next morning.

The chronicler assigned this strange occurrence to the work of the devil, surmising that he made them in "round circles, and plac't every straw with that exactness that it would have taken up above an Age for any Man to perform what he did that one night."

Wondrous markings

In more recent times the study of these intricate, circular marks of unknown origin in crop fields has captured the attention of the public around the world. Modern scientific research into the crop circle phenomenon began in the early 1980's in Britain, when in the summers of 1980 and 1981 circular marks began to appear in fields of wheat in the south and southwest of England. All the circles showed sharp cut-off edges where the flattened stalks of wheat, bent but not broken, began to swirl in a clockwise

direction. In 1983 circles began to appear in clusters of up to five rings. By 1986 circles banded by rings appeared. As the years passed, the marks became increasingly bizarre, creating intricate and often baffling patterns. Many of the circles were spectacular, forming larger, triangular clusters or seeming to be orbited by satellite rings. Others, when seen from the air, seemed like complex

> **One belief is that crop circles arise from the influence of some kind of alien intelligence, and that the circles are the marks left by alien landing craft.**

pictograms of linking lines and circles. There was no convincing explanation as to why the formations of the circles should suddenly have become more intricate, or why an increasing number of formations should appear each year.

As the appearance of bizarre crop circle patterns began to escalate from the late 1980's, so did the many speculations about their origins. Those who believed in paranormal phenomena became increasingly excited, while the scientists grew progressively skeptical.

Alien marks

Many theories have been put forward to explain the origin of crop circles; ranging from the bizarre to the slightly more credible. One belief is that crop circles arise from the influence of some kind of alien intelligence, and that the circles are marks left by alien landing craft.

Dowsers often claim they have obtained strong energy readings when dowsing crop circles. They believe the circles are a manifestation of some psychic disturbance. Yet some dowsers, it appears, have obtained consistently strong results from crop circle systems that have later turned out to be hoaxes.

In addition, some people, who have been influenced by New Age ideas, argue that crop circles are places where psychic healing, heightened levels of consciousness, and an improved sense of well-being can be experienced. Sadly for such claimants, these beneficial energies also appear to be experienced whether or not the circles are in fact real. On a more mundane level some people have suggested that the circles are formed by wild creatures such as rampaging hedgehogs, or other animals.

Aircraft landing

Another explanation is that turbulence from the rotor blades of a helicopter or the depression made by a light aircraft landing might cause the circles. But investigations have shown that while animals and aircraft can certainly create similar depressions in wheat, the actual marks they leave are not consistent with the markings found in crop circles.

In September 1991 two 60-year-old retired English artists, Douglas Bower and David Chorley, leapt into the crop-circle fray. The men said that they had been faking crop circles for years. Bower and Chorley showed the press how they created the hoaxes, using rudimentary tools of planks of wood and string to create repeated, complex, insectlike patterns. They claimed that they had produced their hoax crop circles at night, working from detailed drawings.

Hoaxes exposed

Adding to the complexity of the situation is the fact that members of the scientific body CERES (Circle Effect Research Group), which was set up in 1980 to study crop circles, had spent the summer of 1991 trying to establish which of the 400 circles that had appeared that summer were hoaxes and which were genuine. The CERES group certainly knew that pranksters were rampant and that some had nearly been caught.

Now Bower and Chorley confirmed their suspicions. They claimed that they had created hundreds of circles over a period of 13 years, beginning in 1978. Nevertheless, the two men had operated only over a small area of southern England. Therefore they could not be responsible for the hundreds of circles reported elsewhere, in Britain and the world. Nor could they have made crop circles that appeared before 1978.

Scientists have proposed many theories to explain the appearance of crop circles. Physicists believe they might be the result of electro-magnetism. An electromagnetic field can be created in a laboratory when a current of electricity is passed through a coil of wire, usually copper. The copper coil then attracts ferrous metal. This principle, it appears, can occur in nature, and might be applied to plant stems, but it would require tremendous force to achieve the crop circle formation.

Fairy rings

Botanists have suggested that the fungi that attack grass and mossy ground — to produce what are known in folklore as fairy rings — may provide some explanation. They have investigated to see if such fungi might attack the roots of wheat, causing it to fall. So far research has not confirmed such a connection.

British crop expert and meteorologist Prof. Terence Meaden has perhaps offered one of the most plausible theories. Meaden believes crop circles

Hoax circle
One method of faking a crop circle is for one person to pivot around, pulling against another, while stamping down the corn.

Spectacular patterns
The intricate designs of these interconnecting crop circles, which appeared in July 1991 in the south of England, mark them out as probable hoaxes.

EYEWITNESS EVIDENCE

Among the most reliable data available to crop circle researchers are reliable eyewitness accounts of crop circles forming. The following examples are typical of the accounts researchers are using in their attempts to unravel the enigma of crop circles.

◆ In July 1982 Ray Barnes was taking a country walk near Westbury, Wiltshire, in the southwest of England, when he heard a humming wind and the sound of falling corn. Within three seconds a corn circle had formed in the field beside him. Within the circle that formed, the stalks lay swirled outward from the center. Despite the flattening, the stems of the corn were not broken or kinked: they had simply bent over just above the ground.

◆ In the summer of 1989, as Sandy Reid left for work at dawn one morning near Dundee in Scotland, he heard a commotion in the air. He looked up and watched as a spiral of whirling air caused a crop circle to form in a field nearby.

are formed by an atmospheric phenomenon called a plasma vortex. When an oncoming wind butts against a hill, the downwind airflow can produce violent vortices. When the air is sufficiently humid, the vortices become visible as whirling columns or hollow spheres. As the spiraling vortex of air hits the ground, it flattens whatever crops are beneath it.

Laboratory experiments

Experiments conducted by Prof. Tokio Kikuchi of Kochi University, Tokyo, to create such a vortex under laboratory conditions, have shown that blades of grass can indeed be made to bend — but not, it seems, to the extent that is needed to form a flattened circle.

However, firsthand accounts of people who say they have watched crop circles form add credence to Prof. Meaden's theory. Eyewitnesses consistently report that the wheat fell in a matter of seconds under the effect of what appeared to be local gale-force winds while the air around remained calm. The humming sounds, flashes, and balls of light that witnesses often report might also be consistent with the effects of the strong electrical charges that a vortex might generate.

Swastika pattern

Meaden, however, has had to reassess his theory since the discovery of a crop circle in June 1989, in southwestern England. The wheat in the circle was flattened in four different directions to form a startling swastika pattern. Meaden's vortex theory had

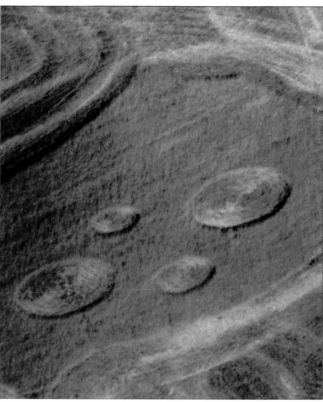

North American rings
Four plain crop circles appeared on September 3, 1991, in Warner, Alberta, Canada. The farmer carefully harvested around the formation.

suggested that in a genuine circle the constituent parts, such as the circle and its rings, swirled in opposite directions.

Crop circle researchers around the world have made certain conclusions: that all complicated circle patterns are probably hoaxes, whereas most of the simpler circles are more likely to be real. Many people who examine genuine circles are impressed by the outward swirling of the wheat, the interwoven layers, and the twisted straws. Yet, as crop specialist John Graham put it, the bending of the stalks is no different from that caused by normal gusts of wind.

An ongoing debate

More research is needed on this puzzling phenomenon. In the meantime, the most rational view has been voiced by one of the world's leading scientists, Prof. Stephen Hawking of Cambridge University, author of the influential *A Short History of Time* (1989). He believes that crop circles are totally explicable by natural atmospheric vortices on the one hand and hoaxing on the other.

Crop circle authority
Meteorologist Prof. Terence Meaden is a leading authority on crop circles. His plasma vortex theory is the most viable scientific explanation to date to explain how crop circles are formed.

CASEBOOK
VIVIENNE AND GARY TOMLINSON

The gusts of wind began to push against the Tomlinsons. They felt the wind coming at them from the side and from above...so great was its force that they were pushed off the path and into the wheatfield.

THURSDAY, MAY 17, 1990, was humid and still. At about 9 P.M. Vivienne and Gary Tomlinson decided to take a late evening stroll along a public footpath that led across a wheatfield close to their home in Hambledon, Surrey, in the south of England. As they were walking along the narrow footpath by the edge of the wheatfield they noticed the wheat to their right, which was dry, light green in color, and two feet high, swaying very gently.

Suddenly, there was a change in the pattern of the wind. It seemed to be coming from two directions at once. In the center of the field, the wind seemed to be gathering force, pushing strong gusts through the wheat. The soft, rustling noise changed now, sounding to Mrs. Tomlinson like the continuous shrilling of a high-pitched flute. The Tomlinsons looked up to see if this sudden change in the wind could be caused by a helicopter overhead. They could see nothing.

Caught in a vortex
The gusts of wind grew stronger and began to push against the Tomlinsons. They felt the wind coming at them from the side and from above, and so great was its force that they were pushed off the path into the wheatfield. Now the wind had completely encircled them. Mrs. Tomlinson could see the swirling air coming at them. It had cloud vapor or mist in it. Looking down, the couple saw that the wheat was being driven down to the ground beneath them.

The Tomlinsons appeared to be trapped in the center of a crop circle as it was forming. It seemed as if the little whirlwinds gathered and increased in size, and as each of them caught up sections of the wheat, the section was laid down gently on the field. The Tomlinsons, battered by the spinning wind, found it increasingly difficult to stand upright. Mr. Tomlinson's hair was standing on end. Mrs. Tomlinson's ears began to ache from the piercing noise. As the number of vortices appeared to increase, Mrs. Tomlinson panicked and with a great effort pulled her husband out of the circle. The formation of the strange circle into which they had been drawn was completed in seconds. After that, the couple watched as the wind split into two currents and zigzagged off into the distance, where it formed other circles in the wheat. The whole episode, including the shaping of the lesser circles, had only taken about seven minutes.

Once the whirling vortex had passed, everything became calm again. The Tomlinsons said they could feel their bodies tingling as an after-effect of the experience.

BATTLES BEYOND TIME

Reliable witnesses have reported hearing the sounds of battles long past. It is as though the noise of battle is being transmitted from the past by some unknown form of energy.

IN AUGUST 1951, two Englishwomen, Mrs. Dorothy N. and her sister-in-law Miss Agnes N., were on vacation at the seaside village of Puys, near Dieppe, in Normandy, France. At about 4 o'clock in the early hours of August 4, the women claimed that they were suddenly awakened by the sound of gunfire, the whine of planes dive-bombing, and men's cries.

Sounds of war
Over the next three hours they listened, appalled by what was later identified as an apparent blow-by-blow replay of a Second World War battle: the Canadian and

> In the early hours, the women claimed that they were awakened by the sound of gunfire, the whine of planes dive-bombing, and men's cries.

British raid on the German-held port of Dieppe. The night-time attack on August 19, 1942, was bloody and unsuccessful, and of the 6,086 Allied troops who participated in the attack, there were 3,623 casualties.

Uncannily similar

The two women — who happened to be staying in a house used by German troops during the war — made detailed notes of what they heard, and from these they later wrote statements describing their experience. They sent their accounts to the Society for Psychical Research, which published a report on them in the May–June 1952 issue of its journal.

In its investigation of the case, which it termed a collective auditory hallucination attributable to psi, (parapsychological mind powers), the society consulted Allied military records. These showed that the women's account of the battle followed closely the timetable of the raid on Dieppe that had taken place nine years earlier, part of it at Puys (see table at right).

No one else heard

The society noted that Mrs. and Miss N. seemed to be entirely well-balanced individuals and had shown "no tendency to add colour to their accounts." The society was also convinced that the women had not simply misinterpreted ordinary nighttime noises. No one else in the vicinity, including a neighbor who slept badly, had heard the sounds of battle, which the women described as being of "amazing" loudness.

It is true that Mrs. and Miss N. owned a guidebook on Dieppe, which contained a detailed account of the 1942 raid. But both women remained insistent that they had not read that account, nor any other description of the military operation, before they underwent their remarkable experience.

The society reached the conclusion: "We think the experience must be rated a genuine psi phenomenon."

Table showing comparison of notes from Mrs. and Miss N.'s account with the timetable compiled from military records of the actual events of August 19, 1942.

WOMEN'S STATEMENT	MILITARY RECORD
About 4 A.M.: Sounds of gunfire, shouts, cries of men above a stormlike noise.	**3.47 A.M.:** German ships and Allied assault vessels exchanged fire. Probably shouting by German defenders on Puys beach.
5.07–5.40 A.M.: After a quarter-hour silence, intense noise began again. Dive-bombers heard.	**5.07 A.M.:** First Allied landing craft beached at Puys and met heavy fire. **5.15 A.M.:** Low-flying Hurricanes attacked German defenses. **5.20 A.M.:** Main landing began at Dieppe, meeting heavy resistance.
5.40–5.50 A.M.: Silence.	**5.40–5.50 A.M.:** Naval bombardment ceased.
5.50 A.M.: Noise of aircraft broke silence.	**5.50 A.M.:** Wave of British planes reached Dieppe.

WHAT KIND OF ENERGY?

Energy is a mysterious concept that scientists have never been able to define satisfactorily. Could there be some forms of energy, which might cause paranormal phenomena, that await discovery?

THE HAND OF A MAN out walking reportedly suddenly bursts into flames. A shower of fish falls from a clear blue sky....Those who believe that strange forces are at work behind such happenings suggest that paranormal forms of energy might be the cause. Yet invoking energy of any kind to explain bizarre phenomena is not as helpful as it might appear. This is because scientists are not fully clear exactly what energy is.

Exertion of power

On a mundane level, we may have little trouble in understanding the concept of energy, defined in Webster's dictionary as: "Vigorous exertion of power...usable power." Take, say, a man cycling uphill. To produce the effort required to push the pedals around, he burns energy obtained partly from the oxygen he inhales, partly from the food he has eaten. The plant constituent of the man's food has gained its energy from the sun, earth, and air; and, in a cyclical process, the air has acquired its oxygen from plants.

Yet it is when we attempt to probe the essence of this energy that problems arise. *The Illustrated Thesaurus of Physics*, published by the Cambridge University Press in England, defines energy as: "The physical quantity present in a field enabling work to be done when the field forces act on a body or particle." If the lay reader inquires: "Physical quantity of what?" he or she will find no answer in the scientific literature. In the field of physics, energy is known only in those specific forms in which it shows itself — for example, electromagnetism, heat, or sound waves — not in itself.

Whatever energy may be generically, it is possible to speculate that science has not discovered all the specific forms that it might take. In this respect, present-day scientists might be, in theory at least, in a similar position to those of the 18th century if they had been called on to explain the workings of television.

Moreover, some of the world's most outstanding scientists have proved to be notoriously poor forecasters in their field. The English physicist and mathematician Lord Kelvin (1824–1907), whose towering achievements included the mathematical analysis of electricity and magnetism and the invention of the absolute-temperature scale, once pronounced: "X-rays will prove to be a hoax." And the scientist who developed the theory of atomic structure, the physicist Lord Rutherford (1871–1937), declared:

> "Anyone who looks for a source of power in the transformation of the atom is talking moonshine."
>
> **Lord Rutherford**

"Anyone who looks for a source of power in the transformation of the atom is talking moonshine."

Whenever, therefore, we hear a scientist dismissing out of hand a theory that has been advanced to account for paranormal phenomena, we might bear in mind that this may be because the theory does not fit in with his or her preconceived ideas.

Essence of energy

Modern physics does contemplate the possibility that, as well as the three dimensions of space and the fourth of time, there may be up to seven more dimensions in the universe. Physicists theorize that these extra dimensions might have been formed at the moment the universe was created and that they exist on an unimaginably small scale. Although these dimensions remain a theoretical possibility, their existence in the universe still remains to be proven. Could it be that these hidden dimensions might explain the occurrence of bizarre phenomena? It seems the answer remains locked within the limits of our current understanding of the universe we live in.

CHAPTER TWO

MYSTERIOUS ANIMALS

Alligators and kangaroos at large in the city, entombed toads, black hell-hounds, lake monsters — reports of these and other animal marvels abound. Endless attempts have been made to explain them away, but not always successfully.

On July 28, 1977, Stan Thompson, a farmer, was relaxing with his wife and friends in McLean County, Illinois, watching radio-controlled model planes diving and darting overhead. Suddenly his wife exclaimed: "Look at that funny-looking plane that guy's flying — it's a bird!" Thompson and his friends turned and saw an enormous bird flying overhead. Its body was as large as a man's, its wingspan 10 feet. It had a long, white-ringed neck and a hooked beak.

Thompson informed the police of this bizarre event. Yet this was not the only time that such a bird was reported in Illinois.

CRYPTOZOOLOGY

The investigation of mysterious animals — out-of-place, unknown, or supposedly extinct creatures — is termed cryptozoology (literal meaning: "the science of hidden animals"). While the professional zoologists of the 19th century busied themselves with classifying the known and the acceptable, amateur naturalists were recording many oddities in the animal kingdom, such as the discovery of living toads entombed in rocks, and the sightings of lake monsters, sea serpents, and other "unbelievable" creatures.

Sea serpents

Among the pioneers of this early cryptozoology was the English naturalist Philip Gosse, author of the popular *Romance of Natural History* (1860–61). Gosse wrote about the possible existence of sea serpents, unicorns, and mysterious great apes.

In modern times, perhaps the outstanding cryptozoologist is the very man who coined the term — the Frenchman Dr. Bernard Heuvelmans. In a series of highly researched, scientifically objective books, beginning with *Sur la Piste des Bêtes Ignorées* (*On the Track of Unknown Animals*) (1955), Heuvelmans has explored this controversial subject in greater depth than any other investigator. And his conclusion is that there may well exist in the world many species of large animals unknown to conventional zoologists.

The Nile monitor lizard
This unmistakable animal lives only in Africa. So how did three large specimens find their way to Florida in 1981?

That same month dozens of other witnesses reported similar sightings. All the descriptions of the giant, mysterious bird closely fitted that of the Andean condor, the world's largest flying bird. The northernmost limit of the condor's range is the southern end of the Colombian Andes, some 2,500 miles away. And so the question became: If the bird was indeed an Andean condor, what was it doing soaring over the lush farmlands of Illinois?

Andean condor

Giant lizards

Florida is a subtropical region, and alligators are not uncommon there. In 1981, within the space of four weeks, three giant reptiles of another kind made a startling appearance in the area. They were identified as Nile monitor lizards, normally found in Africa.

On June 20 James Kilgore, superintendent of the La Mancha golf course in Royal Palm Beach, saw a 6 ½ foot-long Nile monitor crossing the fairway. He lassoed it and took it to the Florida Game Commission. The creature ended up at the local Lion Country Safari.

The following month, on July 12, Donald Wilton, a retired carpenter in Hypoluxo, found a five-foot monitor in his garage. The game commission dispatched its alligator coordinator to capture and remove the beast. Then, only two days later, the game commission was called once

again, this time by a man in North Miami. He had opened the hood of his automobile and nearly fainted when he discovered a five-foot monitor draped across the engine.

Nobody has ever come up with a satisfactory explanation for the appearance in Florida of these creatures. One explanation commonly offered whenever an animal is found in an incongruous location is that it must have escaped from a zoo, circus, or private menagerie. However, inquiries at local zoos and circuses hardly ever turn up any missing animals.

Alligators normally live in tropical or subtropical waters. Yet there are dozens of cases of these reptiles being found and killed in New York. In 1935 Teddy May, superintendent of the New York sewers, became irritated by his inspectors' reports of alligators living beneath the streets — and under his jurisdiction. In addition, he suspected his

> **He had opened the hood of his automobile and nearly fainted when he found a five-foot monitor draped across the engine.**

men of being drunk on duty. One day May went down into the sewers himself to prove that the stories were nonsense. He returned looking shaken. In the smaller pipes of the system, he had seen alligators more than two feet long. Eventually, May got rid of the reptiles by laying down rat poison and by forcing them into the faster-flowing channels of the main pipes, where they were drowned or washed out to sea. Some of the sewermen even took to shooting the alligators with .22 rifles.

Alligators in New York

The problem seemed to be solved — but three years later five alligators were caught in the sewers of New Rochelle,

ALIEN BIG CATS

The eastern cougar was declared extinct in the United States for more than fifty years — yet it is still regularly sighted there.

THE GENERAL CONSENSUS among naturalists is that the eastern cougar *Felis concolor cougar*, which once ranged from the eastern United States to the edge of Alberta and the Plains in Canada, became extinct during the early 20th century. Yet over the past 25 years there have been more than a thousand sightings of the animal. According to John Lutz, who founded what is known as the Eastern Puma Research Network in 1983, *Felis concolor cougar* — a subspecies of the puma *Felis concolor,* common in the western U.S. — has stealthily returned to the eastern states and has colonized the Great Smoky Mountains, the Catskills, the Adirondacks, and the White Mountains of New Hampshire. Lutz describes the eastern cougar as tawny, gray, or black, and smaller than its western cousin — about five feet long (tail included) and 60–125 pounds in weight. He reported that there were 149 sightings of the animal in 1990.

No road deaths?

In Maine the Wildlife Department remains skeptical, though it constantly receives reports of puma sightings. The department points out that most photographs show either a creature that is not a puma or an image that is not clear enough for an identification to be made. The department also notes that if pumas had truly recolonized the East, some would almost certainly have been killed on the roads.

Yet according to professional zoologist Dr. Karl P. N. Shuker, this has indeed been reported. In his book *Mystery Cats of the World* (1989) Dr. Shuker mentions an account of two pumas killed by traffic in South Carolina, one in Georgetown County sometime in 1942–43, and another at Charleston in 1952. In 1971, Dr. Shuker says, a Mr. Buckner shot a puma near Crossville, eastern Tennessee, and arranged for its skeleton to be preserved as a permanent exhibit.

Endangered or extinct?

In 1973, partly because of this specimen, the U.S. Department of the Interior reclassified the eastern cougar as no longer extinct but endangered. Further

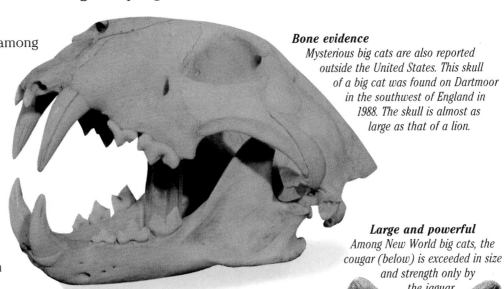

Bone evidence
Mysterious big cats are also reported outside the United States. This skull of a big cat was found on Dartmoor in the southwest of England in 1988. The skull is almost as large as that of a lion.

Large and powerful
Among New World big cats, the cougar (below) is exceeded in size and strength only by the jaguar.

support for its existence came in 1981, when a former director of the U.S. Fish and Wildlife Service, John S. Gottschalk, reported an eastern cougar in the Monongahela National Forest in West Virginia. "It was tawny colored, with the extremely long tail that cougars have," Gottschalk said. "It crouched down, peered at us for five seconds or so, then glided like a wraith into the forest."

Felis concolor is dangerous. In the West, in Colorado, the animal has attacked humans. In 1990 two pumas injured a Boulder County woman, and another attacked two children; in 1991, near Idaho Springs, one fatally mauled a jogger.

Other mysterious big cats, such as black panthers and strange lionlike creatures, have also been reported in the U.S. — often in densely populated areas.

New York. In 1948 there were further sightings of the animals in the New York City sewers, and by the time Thomas Pynchon wrote about sewer alligators in his novel *V* (1963) they had also been seen in the sewers at Yonkers.

One theory for the presence of alligators in the sewers — elaborated upon by Pynchon — is that people buy baby specimens as pets and, when they become too large, flush them down the toilet.

Cold comfort

Alligators in North America can also apparently survive outside the sewer system and — despite their natural need for warmth — in cold weather. One was caught in Hines Park, Detroit, on November 20, 1960.

Alligators — as well as other creatures, such as fishes, frogs, and toads — have been reported in other bizarre circumstances: namely, falling from the sky. On July 2, 1943, for example, the U.S. Weather Bureau reported a fall of small alligators on Anson Street, Charleston, South Carolina, and, in December of the same year, another fall of alligators was reported on a farm in Aiken County, in the same state.

Kangaroos in America

Kangaroos are indigenous to Australia, and, apart from a few captive animals in zoos around the world, are not known to live anywhere else. Yet people from all over the United States have claimed to have seen them. One of the earliest recorded sightings was on June 12, 1899,

---

Despite the great number of kangaroo sightings in the United States over the years, none has been captured.

---

when a cyclone swept through the town of New Richmond, Wisconsin. Amid all the chaos one inhabitant, Mrs. Glover, saw a kangaroo bound through a backyard. Locals blamed the Gollmar Circus, which was in town that week. However, Robert Gollmar, son of the circus owner, later declared that the Gollmar Circus had not possessed a kangaroo at that time. Some 50 years

Urban nightmare
In Thomas Pynchon's novel V *(1963), tourists return to New York from Florida with appealing baby alligators as souvenirs. When the animals begin to resemble their full-grown Florida relatives (like the one shown here), they are discarded, and end up in the sewers. There they start breeding....*

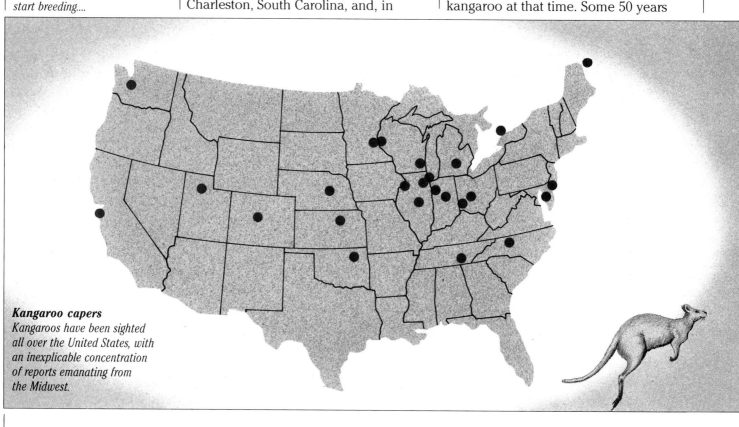

Kangaroo capers
Kangaroos have been sighted all over the United States, with an inexplicable concentration of reports emanating from the Midwest.

The Over Norton crocodile

later, in January 1949, another creature resembling a kangaroo, but apparently capable of jumping on all fours, was spotted near Grove City, Ohio, by Louis Staub, a Greyhound bus driver. In the headlights of his vehicle Staub saw an animal "about 5 1/2 feet high, hairy, and brownish in color. It leaped a barbed wire fence and disappeared."

Kicked in the legs

The 1970's saw a rash of kangaroo sightings in the United States. On the morning of October 18, 1974, two Chicago policemen, Leonard Ciagi and Michael Byrne, responding to a call that a kangaroo was on the loose on the North Side, cornered it in an alley. They decided to handcuff it, so the story goes. Byrne later told a reporter how, when they approached the animal, it began to scream and kick viciously. "My partner got kicked pretty bad in the legs. He [the kangaroo] smacks pretty good, but we got in a few good punches to the head and he must have felt it." Other policemen began to arrive, but Ciagi and Byrne could only watch as the kangaroo bounded down the street.

According to the English periodical *New Statesman* (December 26, 1974), the whole affair was "a healthy joke made up by Chicago policemen and continued by a subtle kind of mass hysteria." However, there was nothing subtle about the headlines in the press the next day, which screamed: KEYSTONE KOPS GO ON KANGAROO KAPER. In the days that followed, there were at least six further sightings of the so-called kangaroo.

Piercing scream

Macropus, the genus to which kangaroos belong, is Latin for "bigfoot." And like Bigfoot — the mysterious apelike creature reported all over the United States — kangaroos can emit piercing screams. Amanda Sutts, of Yardville, New Jersey, who saw a kangaroo-like creature when she was 10, said: "It was about the size of a small calf and weighed about 150 pounds. But the noise is what scared us. It sounded like a woman screaming in an awful lot of agony."

Despite the great number of kangaroo sightings in the United States over the years, none has been captured. Most seem to vanish suddenly. Often the descriptions of the creatures spotted appear to incorporate elements of other animals — or are described as not being kangaroos at all but as creatures that have similar characteristics. In 1975 a hairy, apelike creature that moved like a kangaroo was sighted near the towns of Noxie and Indianola in Oklahoma. It reportedly chased cars and behaved like a Peeping Tom. Search parties were sent after it but it escaped, bounding off at great speed.

ENGLISH CROCODILE

Crocodiles — creatures of tropical climes — are scarcely animals one would ever expect to find in distinctly untropical England. But in 1866 an Oxfordshire landowner, George Wright, claimed that a young crocodile had been found on the Over Norton farm of a tenant, William Phillips. Writing in the *Gentleman's Magazine* of August of that year, Wright described how 10 years earlier the creature had been killed by farm laborers as it emerged from a pile of wood. It had then been stuffed and put in a glass case. The laborers had told Phillips that there were plenty of other young crocodiles in a nearby pond but, upon being asked to produce further specimens, they had failed to do so.

Pickled specimen

Wright sent the specimen to the British Museum, which explained, with an uncharacteristic sense of fantasy, that it had probably fallen in a rain shower (!) or had escaped from a traveling menagerie. Another expert theorized that farm workers had fooled Phillips with a crocodile preserved in spirits.

Kangaroos are indigenous to Australia, and are not known to live anywhere else. Yet people from all over the United States have claimed to have seen them.

ENTOMBED TOADS

Since the Middle Ages there have been hundreds of accounts of frogs, toads, and newts being found alive in sealed-off tombs of stone. The startling implication of some stories is that certain animals may have been incarcerated in stone for centuries.

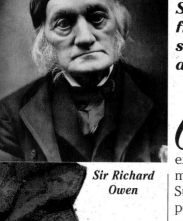

Sir Richard Owen

ONE DAY IN 1982, near the town of Te Kuiti in New Zealand, a gang of railway workers extending a track were trimming mudstone, a type of sedimentary rock. Suddenly one of the men yelled out and pointed to a small cavity he had uncovered in the rock 12 feet below the ground. The others gathered around and saw a live frog inside. Later that day another live frog was found in another hole in the rock. The railway works supervisor, Mr. L. Andrews, stated that it was impossible for the frogs to have fallen into the holes.

The work of the devil
The railway workers' discovery was an example of a phenomenon for which many records exist, dating back to at least medieval times. The 12th-century English chronicler William of Newburgh wrote of entombed toads in his *Historia rerum Anglicarum* (1196–98) and suggested that they were the work of the devil. In the 16th century Ambroise Paré, chief surgeon to Henry III, king of France, gave a first-hand account of finding a toad in rock, and of being told by the quarryman that this was not an uncommon event.

In his *Natural History of Staffordshire* (1686) the English naturalist Dr. Robert Plot described how toads had been discovered sealed up in tree trunks as well as in stone. And the French Academy of Sciences, in its reports for 1719 and 1731, noted the same finding.

Extinct newts
The noted British mineralogist Dr. E. D. Clarke (1769–1822) described, in a lecture he gave at Cambridge University in 1818, the astonishing find made by some workmen while digging a chalk pit. They came across three entombed live newts of an extinct species. Two died shortly afterward, but the third survived for some time.

Toad embedded in stone

During the Victorian era a passion for natural history swept Britain, and entombed toads were prominent among the many discoveries of amateur naturalists. They are mentioned in Mrs. Loudon's work of popular science *Entertaining Naturalist* (1850). Sir Richard Owen, the eminent Victorian paleontologist, was apparently inundated with so many examples of creatures entombed in stone that he asked his wife to cope with them.

In September 1862 one Captain Buckland demanded, in very strong terms, on the letters page of *The Times*, that the organizers of the International Exhibition of that year should get rid of one of their exhibits — the body of a frog and the lump of coal, with frog-shaped indentation, from which it was claimed the animal had been removed.

Buckland expostulated that such an obvious fake lowered the otherwise high tone of the exhibition.

His outburst provoked numerous letters, rebutting his disbelief with seemingly endless accounts of entombed amphibians. These included a toad found in the marble used in the building of Chillingham Castle, Northumberland. Only a few years later, in 1865 — when post-Darwinian scientists were revising traditional theories about the formation of the earth — an amateur geologist, the Reverend Robert Taylor, suggested that, in turning from a gaseous to a solid state, the earth had collapsed inward upon itself, thereby hermetically sealing in the frogs and toads.

Falling newts

In recent years stories have emerged of entombed amphibians found in the permafrost (permanently frozen subsoil) of the Siberian tundra. In 1987 it was reported that one such newt survived for several days, then was placed in the geological museum of Yakutsk, in Siberia. An unnamed scientist from the British Natural History Museum suggested that the newt must have fallen down a crack in the permafrost. This is now a commonly offered explanation for the existence of entombed creatures. However, neither the 19th- nor the 20th-century theory helps to explain why the newts, frogs, and toads should be able to survive for so long in what appears to be a state of suspended animation.

Buried alive

The ability of toads to survive for years in rock has been tested experimentally. In 1825 Dr. William Buckland, professor of mineralogy at Oxford University and dean of Westminster, entombed 24 toads in stone. When he uncovered them a year later, almost half were still alive. The naturalist Philip Gosse believed that more of the toads would have survived if they had been in their torpid hibernating state when buried.

Will Wood, of Eastland, Texas, is said to have sealed a live toad in the wall of a courthouse in 1897, and in 1928 was supposedly present when the building was demolished and the toad rescued, still alive. "Old Rip," as locals named it, lived a year after being released from its prison.

ANACHRONISTIC ARTIFACTS

The rocks of the earth were formed millions of years ago. If the living creatures found entombed in them are a mystery, so are the man-made objects sometimes discovered embedded deep within. Mrs. S. W. Culp of Morrisville, Illinois, was breaking up coal to put on her fire one day in 1891 when she found a gold chain inside one lump. Ten years later, in Uppingham, England, Mr. R. C. Hardman bought a ton of coal and in one piece found a coin reportedly dated 1397. Most of the earth's coal was formed in the late Paleozoic Era, more than 200 million years ago.

Brass in flint

In 1791 near Hamburg, northwest Germany, Herr M. Liesky picked up a flint and, knocking it against another, broke it in two. Embedded in one piece was an ancient brass pin, and in the other was the mould of the pin. Flint is believed to have been formed some 2 million years ago, and archeological evidence indicates that man did not start working metals until as recently as about 1000 B.C.

SIGNS OF THE BEAST

A giant, horned, flying creature emitting a fearsome cry...mysterious hoofprints in the snow running for miles, even over rooftops and walls...terrifying black hounds that vanish into thin air — some have interpreted these phenomena as signs of earthly visitations by the devil himself.

I T WAS 2 A.M. on the snowy night of Sunday January 17, 1909. E. W. Minster, postmaster of Bristol, Pennsylvania, just over the border from New Jersey, could not sleep. He got up, and, as he did so, heard what he told reporters the next day was "an eerie, almost supernatural sound." He looked out over the Delaware River and saw flying diagonally across it a monstrous creature, "emitting a glow like a firefly. Its head resembled that of a ram, with curled horns, and its long thin neck was thrust forward in flight. It had long thin wings and short legs...it uttered its mournful and awful call — a combination of a squawk and a whistle."

Terrifying chase
Hounds of hell hunting the soul, in an illustration by Doris Burton of Sir Harold Boulton's poem "The Chase Wares Hot [sic]" (1926).

Unnerving cry
At about the same time, John McOwen, a Bristol liquor dealer, also heard the same unnerving cry. Looking down from a bedroom window on the Delaware Division canal, he was astonished to see on its bank a large creature that "looked something like an eagle" and "hopped along the tow path." A third observer was patrolman James Sackville, who, as the creature hopped, then flew away from him, fired his revolver at it several times but missed.

Hoofprints in the snow
The next day residents of the area discovered hoofprints in the snow in their yards. Some days earlier, a similar creature had been spotted in Trenton, New Jersey, where it left hoofmarks on roofs. During the following week it was sighted in Woodbury, Swedesboro, and many other New Jersey towns, hoofprints crisscrossing an area of some 48 square miles.

These bizarre goings-on of what has come to be called the Jersey Devil constitute the most concentrated reports of sightings on record. The devil was first allegedly seen by American Indians centuries ago, and since 1909 has been sighted on numerous occasions.

Various explanations for the sightings of 1909 have been proposed. One is that the Jersey Devil was simply a sand hill crane, an extremely large bird with a strident call, formerly a native of New Jersey. Another rather remarkable theory is that the creature was a surviving pteranodon, a relative of the prehistoric pterodactyl. Another explanation is that the rash of sightings was caused by nothing more than mass hysteria. Yet no convincing explanation for the bizarre appearance of the hoofprints has ever been put forward.

Tracks over rooftops
Mysterious hoofprints in the snow also appeared in the county of Devon, England, on the night of 8–9 February, 1855. The townspeople of Topsham, Lympstone, Exmouth, Teignmouth, and Dawlish discovered "a vast number of foot tracks of a most strange and mysterious description" across the snow. The tracks resembled those of a colt's hoof and were an unvarying eight inches apart. They ran for more than 100 miles in a straight line over rooftops, walls, gardens, courtyards, and fields. Moreover, the mystery tracks were not halted by the estuary of the River Exe but continued in the same fashion on the farther side.

Such questionable if intriguing evidence did not prevent the eminent paleontologist Sir Richard Owen from theorizing that the prints were those of a badger. The Reverend G. M. Musgrave, vicar of Withecombe Raleigh, on the other hand, declared the tracks to be those of a kangaroo; better this, he said, than "the dangerous, degrading, and false impression that it was the devil."

Hounds of hell
Mysterious footprints in the snow are also seen each year traversing a small hill in the region of Galicia near the Polish–Ukrainian border; such bizarre footprints are interpreted by the local people to be supernatural in origin.

> **Supernatural black dogs are commonly described, by those who claim to have seen them, as large and terrifying, with blazing eyes and sometimes a luminous outline.**

Supernatural black dogs feature strongly in European folklore, in which they are sometimes called "hounds of hell" — manifestations of the devil. The dogs are commonly described, by those who claim to have seen them, as large and terrifying, with blazing eyes and sometimes a luminous outline. A man in Somerset, England, who supposedly encountered one in 1907, spoke of its "great fiery eyes as big as saucers."

Fictional hound
It was undoubtedly one of these creatures that Arthur Conan Doyle had in mind when he wrote *The Hound of the Baskervilles* (1902). In this Sherlock Holmes tale the hound is described as: "a foul thing, a great black beast...larger than any hound that mortal eye has ever rested upon...its eyes glowed with a smouldering glare, its muzzle and hackles and dewlap were outlined in flickering flame."

In some encounters with black dogs, apparently, the beast suddenly vanishes before the beholder's eyes. Sometimes, this happens in a flash of light or puff of sulfurous smoke; sometimes the dog shrinks or expands and slowly becomes transparent; and sometimes it turns into another creature, generally human.

Fearsome black dog
Most reported encounters with black dogs have taken place in lanes and ancient trackways near water, along leys (straight lines that run through a series of landmarks and ancient sites, such as crossroads and churches). For example, on the morning of August 4, 1577, during a flash storm in Bungay, a town in Suffolk, England, a fearsome black dog was said to have entered the parish church and run through the congregation. As it passed between two people kneeling in prayer, the story goes, they were struck dead. Another person was brushed by the dog, and lived to tell the tale, but shriveled up and withered.

Unholy numberplate
Harry Moore, an Australian priest, chose 666 as the license number of the automobile he drove around Norfolk Island in the south Pacific, in the late 1980's. This was apparently an attempt to rid the number of its satanic association.

THE DEVIL'S NUMBER
According to the bible, 666 is the number of the ten-horned, seven-headed beast that uttered "haughty and blasphemous words" against God (Revelation 13: 5, 18). The English black magician Aleister Crowley (1875–1947), who called himself the Great Beast, adopted the number as his mark. Such satanic association exists even today and has been acknowledged by a governmental organization: in 1991 the Driver and Vehicle Licensing Agency in Britain announced that 666 was not to be used on any future vehicle registration.

ANIMALS AND MAN

There are many amazing stories of animals appearing to help man or act like him. Do such accounts signify truly humanlike behavior in these creatures, or are they simply evidence of an apparent need in man to project his own characteristics onto the animal world?

IN 1626 A THEOLOGICAL WORK by John Frith was reprinted in London with a new title, *Vox piscis* [*Voice of the Fish*]: *or, The Book-fish*. The reason for this bizarre title was that, on June 23 of that year, a copy of the original edition had been found, Jonah-like, in the belly of a codfish. Mr. Joseph Mead, of Christ's College, Cambridge, had been wandering through the town's market when he became aware of a hubbub at a fish stall. There he saw a newly gutted cod and, beside it, covered with fish guts, a sailcloth-bound copy of John Frith's work, which it was claimed had been taken from the cod's stomach. Mr. Mead took possession of the book and arranged for it to be reprinted under the new title.

Helpful fish

Stories of fish rendering assistance to humankind are quite common in Greek and Celtic mythology and in medieval legend. In *Gesta pontificum* (*Acts of the Bishops*) (1123), the English historian William of

---

As the ship neared Italy, a large fish leaped onto the deck and, upon being cut open, was found to contain the key to Egwin's fetters.

Malmesbury tells the story of how a fish dramatically intervened in the life of Egwin, who was bishop of Worcester from A.D. 692 to 717 and was later canonized as St. Egwin. Having become embroiled in disputes in his diocese, Egwin decided to make a pilgrimage to Rome. Before sailing, he determined to expiate the sins of his youth by fettering himself. He then threw the key into the River Avon. As the ship neared Italy, a large fish leaped onto the deck and, upon being cut open, was found to contain the key to his fetters. Egwin took this as a sign that he should release himself.

Friendly dolphins

Dolphins are renowned for their friendliness to humans. In Herodotus, Aesop, the Plinys, and other classical literature, there are many accounts of people hitching a ride on dolphins or being rescued by them. At Laguna,

WEIRD WILDLIFE

Islands ought to be unlikely places for the harboring of out-of-place land animals: the sea should prevent them straying beyond their natural range. But in Britain many such animals have been sighted, including baboons, boars, and bears.

Mysterious immigrant
The Chinese muntjac deer has become established in the southwest of England, but no one knows how it arrived there.

WHY ARE MONGOLIAN GERBILS and African clawed toads living on the Isle of Wight? What are Himalayan porcupines doing in Devon? Perhaps these and other out-of-place creatures have been brought into Britain — unwittingly or deliberately — by travelers returning from abroad and, escaping or set free, have become naturalized. But this does not explain the presence of those sizable animals, such as baboons, boars, and bears, that have also been sighted. It may be that they are escapees from zoos, circuses, or private animal collections — but these scarcely ever report missing such animals. Whatever the explanation, Britain has a far more exotic wildlife than people realize.

◆ In 1979, in the area around Brassknocker Hill, south of Bath, Avon, oaks and other trees showed the depredations of a creature that came to be known as the Beast of Bath: it stripped off bark, leaving teeth marks 10–20 times larger than those of squirrels. Three men who claimed almost to have caught the beast were convinced that it was a fully grown chimpanzee. Another witness was also of the same opinion, but his companion believed the animal to be a baboon. There was also subsequent speculation that it might be a spectacled bear (a South American bear so named because of the whitish patches round its eyes).

◆ A bear was sighted by a number of motorists on a busy road near Thetford, Norfolk, on June 11, 1979. From the descriptions given, experts decided that the animal was probably a small Himalayan or Malaysian bear.

Wild boar

◆ Four wild boars — long considered extinct in Britain — were seen in Hampshire in the summer of 1972. Two were tracked by dogs but escaped; another, a 200-pound specimen that was eating young trees, was captured; and the fourth, found eating barley, was shot by a farmer. On July 6, 1975, another was sighted on a highway near Sandbach, Cheshire.

◆ In 1976 a baboon was chased by policemen through gardens in Windsor, Berkshire.

◆ In 1979 a band of monkeys was blamed for a series of nocturnal raids on dustbins and greenhouses around Exton, Leicestershire. A local bartender reported losing a capuchin monkey, but the animals sighted were apparently much larger — about the size of spaniels. Monkeys were later reported swinging through the trees at Stamford, Lincolnshire, only a few miles away.

◆ Other strange British colonists include alpine newts, found in Surrey; Chinese and Indian muntjac (species of deer), in southwest England; Chinese water deer, in central and southern England; American catfish in Lancashire, Middlesex, and Ayrshire; cichlids (tropical, mainly African, freshwater fish) and guppies (West Indian freshwater fish), in Lancashire; coypus (South American rodents), in East Anglia; New Zealand stick insects and Japanese sea squirts, in Cornwall; and many more.

in southern Brazil, dolphins and men have been working together, to their mutual benefit, since the mid-19th century. Fishermen wanting to catch red mullet wait on shore for dolphins to give them a signal — a roll in the water — that a school is approaching. They then cast their nets. The dolphins drive the school toward the nets, which break it up and so make it easier for the dolphins to catch their share of the mullet.

Experiments in the United States and in Holland in the 1980's have established that dolphins can follow up to 700 instructions, possess a well-developed memory, and can even understand certain rules of syntax. And many scientists who have studied the intricate songs of the humpback whale believe that they may form a comparatively advanced means of communication.

The Elberfeld horses

Marine animals are not the only creatures that man has suggested are capable of performing remarkable feats. At the beginning of the century the Belgian writer Maurice Maeterlinck (1862–1949), winner of the Nobel Prize for literature, published a remarkable essay in the New York *Metropolitan Magazine*. Entitled "The Elberfeld Horses," it recounted how Maeterlinck had investigated the claims of a Herr Krall of Elberfeld (now part of Wuppertal), in Germany, that he owned four "intelligent" horses, which could spell and perform other mental feats.

Striking lucky?

Krall explained to his visitor that he had taught the horses a system whereby they could communicate each letter and each numeral by striking a board a certain number of times with the hooves. Left alone with one of the animals, Maeterlinck repeated over and over to it the name of the hotel where he was staying: Weidenhof. Then, wrote Maeterlinck, the horse "straightway blithely raps out the following letters, which I write down on the blackboard as they come: WEIDNHOZ."

Krall returned and told the horse that he should have ended the word with an F, not a Z. The animal obliged by rapping out an F. The Belgian writer said that he then saw other horses solve mathematical problems (including the extraction of the fourth root of a six-figure number), identify pictures, and distinguish colors. "I have made a point of stating only what I saw with my own eyes..." he declared, "with the same scrupulous accuracy as though I were reporting a criminal trial in which a man's life depended on my evidence." Maeterlinck concluded that the horses may have been using intelligence of an extent never credited to them before, and possibly some kind of psychic power "similar to that which is hidden beneath the veil of our reason."

Talking animals

The ability of certain birds to imitate the language of man is well known. Sparkie, a British budgerigar who died in 1962, could recite eight nursery rhymes consecutively. No doubt he learned the rhymes by rote, without knowledge of the meaning of the words. But could some talking animals really understand what they are saying? The results of long-term tests on a talking 15-year-old African gray parrot named Alex are impressive. The tests, begun in 1977, were carried out by Dr. Irene Pepperberg, an ethnologist at the University of Arizona. The parrot was shown seven objects in different colors and materials, including a green block of wood. When asked which item was green, Alex immediately replied: "Wood." Given 48 questions on similar problems, the parrot answered 76 percent of them correctly.

Hoover the harbor seal

Those who heard Hoover, a trained harbor seal, talking in 1981 at the New England Aquarium, Boston, also wondered whether perhaps the animal understood what it was saying. It would surface, for example, call out: "Hello there," "How are you?" "Come over here," or "Get out of here," and then laugh heartily. A bemused Ros Ridgway, the aquarium's publicity director, commented: "Harbor seals are not supposed to make these sounds."

Dolphin ride
Taras, son of Poseidon, Greek god of the sea, is pictured here on a Greek coin (c. 5th century B.C.). He is seen riding on the dolphin that, according to Greek myth, rescued him from a shipwreck.

Alex the parrot was shown seven objects in different colors and materials, including a green block of wood. When asked which item was green, Alex immediately replied: "Wood."

ANIMAL HOAXES?

A prehistoric apeman in Minnesota, a talking mongoose on the Isle of Man — however unlikely these animals appeared, many people, some of them quite eminent, were prepared to consider them seriously.

Phantom footprint?
A paw impression claimed to be from a talking mongoose; it was proved to be a fake.

Heuvelmans came to the conclusion that the creature was a recently living specimen of a Neanderthal man, a subspecies of *Homo sapiens*, that had become extinct about 35,000 years ago.

O N DECEMBER 17, 1968, two eminent cryptozoologists, the Frenchman Dr. Bernard Heuvelmans and his friend the American Ivan T. Sanderson, gazed wonderingly at what lay encased in ice in a glass-topped freezer box. Its body covered in long brown hair, yet with the face of a primitive human being, it seemed to be some kind of "apeman."

The owner of this strange creature, which came to be known as the Minnesota iceman, was a showman named Frank Hansen. Hansen exhibited the iceman at fairgrounds, claiming that it was the "missing link" between humans and apes. A snake dealer who had seen the exhibit contacted Sanderson, knowing that, as an expert on Bigfoot, he would be interested.

Photographs and drawings

For three days, inside the trailer where the freezer was housed, Heuvelmans and Sanderson took many photographs and made many drawings of the iceman. It was 5 feet 10 inches tall and had a hairless face with a prominent brow, an upturned nose, a thin mouth, and a large jaw. The creature had a short, muscular trunk; a hairless groin; short legs, and broad, humanlike feet. The left arm appeared to be broken, one eye was missing, the other hung from its socket, and the back of the head had been severely injured.

Neanderthal man

Hansen was self-contradictory in explaining where the iceman had come from, claiming at one time that he had bought the creature from a dealer in the Far East, at another that it belonged to a wealthy Californian, and on a later occasion that he had shot it himself. Far from concerned with all this, Heuvelmans came to the conclusion that the creature was

Primate poser
A line-drawing of the Minnesota iceman by Ivan T. Sanderson.

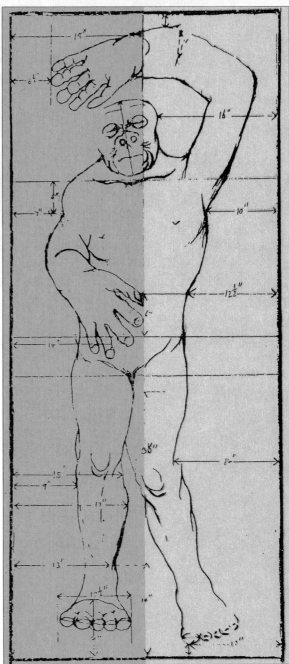

a recently living specimen of a Neanderthal man, a subspecies of *Homo sapiens*, that had become extinct about 35,000 years ago. He named the iceman *Homo pongoides* (the pongids are the manlike apes) and sent a scientific report on it to the heads of various institutions, including Dr. John Napier of the Smithsonian Institution in Washington D.C. Fascinated, Napier informed Hansen in March 1969 that the iceman should be investigated. But Hansen then disappeared.

During his absence, it was claimed anonymously that the iceman had been manufactured in Hollywood by a model-maker named Pete Corrall, a story that Hansen on his return confirmed. (A later account confused the issue by suggesting that it was made by Howard Ball, a Disneyland artist.) Three years later Dr. Napier stated that the presentation of the iceman as a primitive man had been a brilliant hoax. Heuvelmans, however, never accepted that the iceman was not genuine.

Talking mongoose

If the iceman was indeed a hoax, this was undoubtedly perpetrated for publicity and commercial gain. Rather different were the probable motives behind what was almost certainly another animal hoax, which took place in 1931 — the case of the talking mongoose. This amazing creature was said to inhabit an isolated, windswept farmhouse on the Isle of Man, a dependency of Britain in the Irish Sea. James Irving, the 60-year-old head of the household, told reporters that Gef, as the family called the mongoose, sang songs, gave orders, bounced a ball, and in return demanded candies. Irving's story was backed up by his wife Margaret and 13-year-old daughter Voirrey.

Uncanny voice

According to Irving, Gef had a yellow body and brown tail and lurked behind the wooden paneling of the farmhouse. But reporters who visited the house saw nothing of the animal. They simply heard what the man from the *Manchester Daily Despatch* described as the mongoose's "piercing and uncanny voice," as it

conversed with Mrs. Irving. In an adjoining room sat Voirrey. The *Despatch* reporter noted: "When I edged my way into the room the voice ceased."

In 1935, after the Irvings had produced alleged samples of Gef's fur, the celebrated psychical investigator Harry Price gave them to Martin Duncan, of the Zoological Society of London. Duncan reported that they probably came from a long-haired dog. In July of that year Price began a three-day visit to the farmhouse with R. S. Lambert, editor of the magazine *The Listener*. They saw and heard nothing of Gef, who, Irving said, had vanished. They did, however, take away some hairs from Voirrey's dog, Mona. Duncan found these to be identical with the "mongoose's fur."

James Irving then sent Price three paw prints made in modeling clay by Gef. Price gave photographs of these to R. I. Pocock, of the Natural History Museum, who commented: "Most certainly none of them was made by a mongoose."

Joke or hoax?

That was more or less the end of the affair. And eventually the Irvings moved away from the farmhouse, returning to their former obscurity. It seems obvious now that the story has a simple explanation. The high-pitched voice was Voirrey's. In that lonely farmhouse the girl had no other children for company, no neighbors, not even a radio. What could be more natural than that she should create her own imaginary companion? As for Mr. and Mrs. Irving, they probably saw their daughter's fantasy as a joke at first. However, once they had publicly testified to Gef's existence, they no doubt felt committed to the hoax and were prepared to go out of their way to continue it, in order to save face.

Objective eye
The noted psychical investigator Harry Price enlisted scientific help in examining the claims of the Irving family.

Lonely girl
Gef may have been dreamed up by Voirrey Irving, the lonely, imaginative girl who lived in a desolate farmhouse, in order to keep her company. Voirrey was an avid reader of animal stories and may well have based "Gef" on Rikki-tikki-tavi, the clever mongoose in Rudyard Kipling's The Jungle Book.

Monster or hoax?
This remarkably clear photograph of the Loch Ness monster was taken on May 21, 1977, by Anthony Shiels. Because Shiels was a professional showman, his photograph was widely regarded as a hoax.

Nessie a whale?
The streamlined, ferocious killer whale, seen here in Johnstone Strait, British Columbia, Canada, is one of the few known, non-extinct animals large enough to account for the many sightings of Nessie in the loch.

NESSIE AND CHAMP

Are the many accounts of monsters seen in Loch Ness in Scotland and Lake Champlain merely over-imaginative sightings of natural objects? Or do giant creatures, perhaps prehistoric in nature, live in the depths of these and other lakes?

*I*T WAS AN APRIL MORNING IN 1932. Col. L. McP. Fordyce was driving with his wife along the southern side of Loch Ness in the Highlands of Scotland when they were startled to see, about 150 yards ahead, an enormous, lumbering animal coming out of the woods on their left and crossing the road toward the loch. In the account that Colonel Fordyce eventually published in 1990 (he had constantly deferred publication for fear of being accused of perpetrating a hoax), he described the monster in detail: "It had the gait of an elephant, but looked like a cross between a very large horse and a camel, with a hump on its back and a small head on a large neck."

Land monster
Such land encounters form the majority of the infrequent, pre-1933 sightings of the Loch Ness monster, or Nessie, as she is popularly known. Since 1933 almost all sightings have been of Nessie in the loch itself. If the monster does exist, could it be that in 1933 some circumstance drove it out of the thick woods surrounding the loch and into the water? Or is — or was — the land monster a different creature from the one in the loch? If, on the other hand, Nessie is only a figment of the imagination, is it simply the publicity about a lake monster from 1933 onward that has been responsible for almost all subsequent sightings being of Nessie in the loch?

Gas-filled mat
Since 1933, theories about Nessie's identity have ranged extraordinarily wide. They include suggestions that the monster is a killer whale, a huge newt, a family of otters, a gas-filled mat of rotting vegetation, a giant worm, swimming deer, an

Nessie model?
This is the most famous photograph of Nessie. It was taken in 1934 by London surgeon R. K. Wilson. It shows a long-necked, small-headed animal that has become the archetype for most alleged sightings of the monster in the years since.

extinct elephant-like creature, or even a large floating log.

One of the most persistent theories is that Nessie and other lake monsters are surviving plesiosaurs, large marine reptiles that, according to the evidence of fossils, became extinct at the end of the Cretaceous Period, about 65 million years ago. Plesiosaurs had a small head, long, snakelike neck, and large flippers, and they match quite closely the creatures in some alleged photographs and descriptions of Nessie. Reptiles, however, are cold-blooded animals that live in warm conditions. The temperature of Loch Ness's murky waters, where light scarcely penetrates, is a chilly 42°F (14°C), and unless surviving plesiosaurs had somehow evolved over the millennia into atypically warm-blooded reptiles, they could not have survived in the depths of the loch.

Sonar sweep

In 1987 a flotilla of 24 boats, extending across the width of Loch Ness, swept its entire length several times with a sonar curtain (sound waves that are reflected back from any objects they meet). Apart from one reading, of an unidentified large object at great depth, the results, so eagerly awaited, proved disappointing. Loch Ness had refused to yield its secrets.

Famous photographs

The most celebrated of all American lake monsters, Champ, the reputed inhabitant of Lake Champlain, the waterway between New York and Vermont, has been sighted on innumerable occasions since the early 19th century. The most famous alleged photographs of Champ, taken in July 1977 by Sandra Mansi and reproduced in *The New York Times* four years later, show a creature remarkably like that in the most famous photograph of Nessie, taken in 1934. Both are small-headed and long-necked, like a plesiosaur.

Some researchers, such as American biochemist Dr. Roy Mackal, believe that one or more forms of giant prehistoric creature may indeed have survived in the landlocked depths of various lakes across the globe. But so far the evidence consists almost entirely of individual sightings and inconclusive photographs. Until more concrete evidence emerges, such as a monster body, Nessie, Champ, and all the other lake monsters of the world will remain a mystery.

Prehistoric reptile
This model of Elasmosaurus, *a long-necked form of the plesiosaur, a large prehistoric marine reptile, was reconstructed on the basis of fossil evidence. It bears a marked resemblance to various lake monsters allegedly depicted in photographs.*

Champ observed?
In July 1977 Sandra Mansi took this famous photograph of the Lake Champlain monster, which she described as being larger than two horses. The negative was lost, which has led to speculation that the photograph may be a hoax.

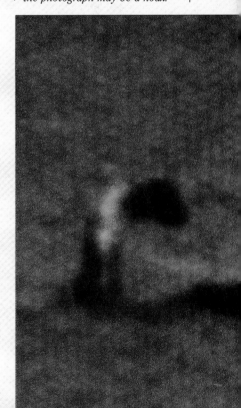

CHAPTER THREE

EXPECT THE UNEXPECTED

Every day somebody, somewhere, is assailed by a strange event or a bizarre coincidence. No matter how experts try, some of these synchronicities are too strange to explain. Take the following case of separated twins, for example.

On August 19, 1939, twin boys were born to an unmarried immigrant at Memorial Hospital, in Piqua, Ohio, and immediately put up for adoption. Later in August 1939, the Springer family of Dayton, Ohio, adopted one of the baby boys. His twin brother, they were told, had died. In the meantime, 80 miles away in Lima, Ohio, the Lewis family adopted a baby boy. They were also told that the infant's twin brother had died. However, six years later the Lewis family accidentally discovered that their adopted son's twin, whom they had believed was dead, was alive.

Twin tests
In 1979 Thomas Bouchard, professor of psychology at the University of Minnesota, put the "Jim twins" through a series of tests that revealed startling similarities in their lives.

ACCIDENTS OF NATURE
There are an estimated 100 million pairs of twins in the world. Twins are defined by the *Encyclopaedia Britannica* as the simultaneous birth from one mother of two live offspring. In the animal kingdom, multiple births are common, but human twins are accidents of nature. (A woman's womb is constructed to carry one child.)

Fraternal twins
Twins are conceived in two possible ways. Fraternal, or dizygotic, twins are the result of two separate ova being fertilized by two separate sperm at the same time. Each twin is carried in the womb in its own placenta. Even though they are conceived and born at the same time, fraternal twins are genetically only as near, and as distant, as any other brother or sister.

Identical in every way
Identical, or monozygotic, twins are the product of a single ovum, fertilized by one sperm, then splitting in two. The resulting embryos are always the same sex and have the same inherited genetic makeup. They also share a placenta which makes competition for food and oxygen intense.

Thus James Springer grew up believing that his twin was dead, while James Lewis grew up not knowing where his twin was. But in 1977 James Lewis decided to track down his twin, and the brothers were finally reunited in 1979 when they were 39 years old. "I looked into his eyes and saw a reflection of myself," James Springer recalled. At their first meeting the brothers hugged, then burst into tears.

The "Jim twins"
The lives of the "Jim twins," as they came to be known, revealed a startling catalog of coincidences. Each had an adopted brother called Larry, and as boys both had a dog named Troy. At school both men had enjoyed math but disliked spelling. Each twin had married and divorced a woman named Linda, and then remarried a woman named Betty. Each man had a son: they were called James Alan and James Allan. Each of the twins had regularly taken their annual vacation at the same resort in Florida. Both men had worked as gas station attendants and for the same burger chain.

The twins had identical smoking and drinking habits, bit their nails compulsively, and had

> ## "I looked into his eyes and saw a reflection of myself."
> **James Springer**

suffered from migraines at the same time in their lives. Another coincidence was that the six-foot-tall brothers had gained 10 pounds extra weight, and lost it again, at the same time in their adult lives.

Inherited characteristics
The astonishing coincidences in the lives of Springer and Lewis prompted psychologist Thomas Bouchard to study twins in an effort to discover how much they can tell us about human behavior, and the relationship between inherited characteristics (acquired genetically) and environmental influences. These influences form the two parts of what is known as the nature-nurture equation.

Coincidences such as those seen in the Jim twins' lives have led some people to believe that there may in fact be a psychic connection between identical twins. Such thinking appears to have inspired the celebrated 19th-century French writer Alexandre Dumas in his novel *The Corsican Brothers.* Dumas tells the tale of identical twin boys who are separated at birth. The boys are reared in separate countries, and grow up in complete ignorance of each other's existence. Yet their lives are punctuated by inexplicable feelings of joy and pain which they cannot understand until they

discover each other's existence. The story ends in tragedy when one brother is fatally wounded in a duel. At that moment his brother also experiences the pain of the bullet in the same spot. He also dies.

Sisterly feeling

Our everyday world may not be as romantic as the world of fiction, yet it can also display extraordinary synchronicities. Mrs. Joyce Crominski from Australia, for example, described how one night her sister Helen, age 19, woke at 11:15 P.M. screaming from excruciating chest pains. Helen's parents sent for an ambulance, but the teenager died on her way to the hospital. That very same evening, Helen's twin sister, Peg, also died in an ambulance. She had had a car accident in which the steering wheel of the car she was driving smashed into her chest. This reportedly happened at the very time that Helen awoke screaming in pain.

Coincidences between twins even appear to cross from one generation to the next. In Leicester, in the north of England, in October 1988, identical twins Susan Harcourt and Fay Johnston gave birth within six minutes of each other at the city's Royal Infirmary. Susan's baby was not due for another six weeks. Twelve years previously Susan had been on vacation when she was suddenly rushed to the hospital suffering from severe stomach pains. However, Susan's

symptoms disappeared when she found out that they had started at the same time her sister, Fay, had gone into labor.

From birth to death

As twins enter the world together so may they depart from it. Identical twins Ruth Standon and Rachel Garber were born on June 13, 1913. Both sisters died of heart failure within four hours of each other, on May 8, 1988. Ruth died in Dayton, Ohio, while Rachel died, less than 40 miles away, in Greenville, Ohio.

When the police in Barnoldswick, Lancashire, England, went to tell 86-year-old Lizzie King that her twin sister, Edith — who lived just two houses away on the same street — had collapsed behind her kitchen door and died, they found Lizzie dead, too. She was also lying dead behind her kitchen door. The coroner's report showed that both sisters had died of heart failure at about 9:45 A.M. on the same day in April 1988.

Hollywood twins
In the 1941 Hollywood movie,
The Corsican Brothers, both roles
of the identical twins, Louis de
Franchi and his brother, Lucien,
were played by the swashbuckling
Douglas Fairbanks, Jr.

Twin power
Twinsburg, Ohio, has an annual
festival of twins. In 1988 a record
3,618 pairs of twins gathered in the
town from all over the world.

Twin doctors
Marguerite and Mary Engler are identical twins. They are both researchers into the control of heart disease at the University of California School of Nursing in San Francisco. The Engler twins are immediately recognizable as identical, and their similar academic achievements reflect a comparable intellectual ability.

Yet twins are not the only people related by blood who are on the receiving end of uncanny family coincidences. In July 1987, Will Hewitt of Southport, England, went to his local library to consult the list of registered voters. A woman was looking through the list when he arrived. They started to talk, and the woman introduced herself as Vivien Fletoridis from Australia. She explained that she was trying to track down a man named Hewitt who had had a wartime romance with her mother. Vivien was the result of this liaison and she had been put up for adoption soon after she was born. Her adoptive parents had taken her to Australia when they emigrated there in 1954. Vivien had returned to England to trace her natural father. Will Hewitt told her that she need look no further.

Mother reunited
In 1991 an astonishing reunion took place at the Stop-in-Store in the town of Roanoke, Virginia. Tammy Harris knew she had been adopted at the age of two because of her mother's alcoholism. When she reached 21, in March 1990, she decided she wanted to trace her natural mother. The following February, Tammy discussed this problem with Mrs. Joyce Schultz who, six months before, had started to work with her at the checkout in the Stop-in-Store. When Tammy showed her fellow worker her birth certificate, an overjoyed Mrs. Schultz realized that Tammy was the daughter for whom she had been searching for nearly 20 years.

Blind fate
Families who are linked by adoption, not blood, have experienced astounding coincidences. In 1991 a blind 87-year-old army veteran, Ray Ackroyd, wanted to find his two adopted sons, Rod and Shankhill Davis, before he died. Ackroyd and his late wife, Mary, had adopted the boys in 1942 when his regiment was stationed in Wales. But he had lost touch with them for the last 30 years. It appears that Ackroyd had saved money from his weekly pension to make the 250-mile journey by taxi from his home in County Durham in the northwest of England, to Aberaeron near Aberystwyth in Wales, which was the last place he had heard that his sons were living.

This probably seemed like a wild goose chase to the taxi driver, especially as the surname Davis, or Davies, is very common in Wales. When they reached Aberaeron, the taxi driver flagged down a passing driver to ask if he knew a man called Rod Davis. The man said "I am Rod Davis." He turned out to be Ackroyd's adopted son.

Multiple birthdays
In some families birth dates are repeated with an almost astonishing regularity. When, for example, Chadwick Parker Hume was born in Dallas, Texas, on January 7, 1987, he was continuing a family tradition reaching back to his great-grandfather, Lee Royal Jenney, who was born on the same day in 1882. Jenney's daughter Florence was born in New York City on January 7, 1925. Her son Craig Hume, Chadwick's father, was born on the same day in 1955 in Boston, Massachusetts.

Nine times lucky
Bakr Bin Muhammad Hawsawi from Saudi Arabia has maintained a similar tradition. Each of his nine children has been born on the same day in the Arab month of Jamad-al-Thani, worked out from the lunar calendar. However, such a coincidence has its drawbacks. The world of officialdom always refuses to accept Mr. Hawsawi's story, and he has to produce nine birth certificates to prove it.

> **Each of Mr. Hawsawi's nine children has been born on the same day. The world of officialdom always refuses to accept his story, and he has to produce nine birth certificates to prove it.**

THE MINNESOTA PROGRAM

Research into the lives of identical twins who are reared apart suggests that not only is the way they look but also the way they behave strongly influenced by genetic inheritance.

IN 1980 TONY MILAI, who had been adopted by a rich New York Italian Catholic family, met his identical twin Roger Brooks, who had been raised by a single Jewish mother of modest means in Florida. They soon discovered that they had many more similarities than simply appearing to be physically identical. Many of their mannerisms were the same: they held a coffee cup without touching its handle in exactly the same way; they unpacked a suitcase in an identical fashion; and they even used the same brand of imported Swedish toothpaste.

Such similarities between identical twins who are reared apart prompted Thomas Bouchard, professor of psychology at the University of Minnesota, to set up in 1979 a research program focusing on twins and their behavior. He wanted to study coincidences in the lives of identical twins and to see what they revealed concerning the effect that genetic and environmental factors have on our lives. Bouchard believed he would find some human traits that were inherited and others that were clearly environmental in origin.

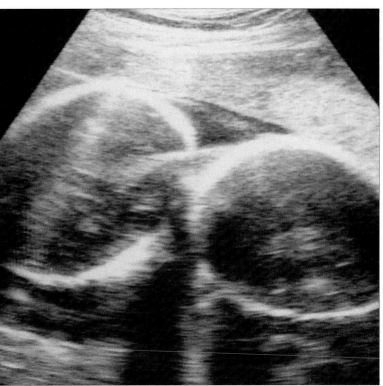

Life in the shared womb
An ultrasound scan shows the prenatal environment of identical twins.

Twin study

The program began by studying 20 sets of identical twins. Each twin was put through a rigorous six days of tests, answering over 15,000 searching questions ranging from family history to childhood environment, education, esthetic appreciations, and color preferences. As might be expected in the study of identical twins, who are genetic clones of each other, the physical characteristics were nearly identical between

Irene Reid and Jeanette Hamilton
Reunited for the first time after 33 years, the Reid/Hamilton twins found they had the same phobias about water and heights.

the sets of twins. In addition, Bouchard found that the sets of twins' brainwave patterns and IQ's had a high level of correlation. However, the psychological profiles of the twins were also much alike. One set of twins, for

example, who had been reared in very different environments — one was a manual worker and the other was college educated and highly cultured — both turned out to have the same ability as raconteurs even though their lifestyles were different.

The psychiatric history of twins also showed similarities. Twins reared in different environments — one in a very strict household and the other by a very tolerant and loving mother — both showed the same patterns of hypochondriacal and neurotic behavior.

Since the start of the program, Bouchard's team has studied 100 pairs of identical twins. Among his conclusions, he asserted that the lives of twins can show us that "most human attributes have at least a small but significant genetic component."

UNEXPECTED WORDS

When in a trance or under hypnosis or while meditating, some people have reportedly started speaking in a language or tongue unknown to them. This bizarre phenomenon has proved difficult to explain.

On MAY 10, 1970, DOLORES JAY was hypnotized by her husband, Carroll Jay, a minister from Elkton, Virginia, in an attempt to cure her backache. When the question "Does your back hurt?" was posed to her, she replied *nein* (German for "no"). Intrigued, he hypnotized his wife again three days later and began probing to see if there was a straightforward explanation for her response in German.

"Trance" personality

Over several months of intensive questioning, a "trance" personality reportedly emerged who identified herself as Gretchen Gottlieb. She claimed that she was the daughter of the mayor of Eberswalde in Germany. In imperfect German, "Gretchen," via Dolores Jay, revealed details of the life that she had lived in the late 19th century. She apparently had died when she was 16 years old. When she came out of hypnosis, Dolores Jay could hardly believe the tape recordings of herself speaking such proficient German. She had never studied the language or known any speakers of German.

Background investigated

Over the next four years an army of researchers questioned, taped, and analyzed Gretchen's responses. They concluded that Dolores Jay's German could not have been learned by simply reading the language. Noted parapsychologist Prof. Ian Stevenson, of the University of Virginia, questioned the Jays closely, and painstakingly investigated Dolores Jay's background and childhood in an effort to uncover any German lessons or links with the language. He found none. A lie detector test seemed to confirm Dolores's claims.

While there are many examples of individuals who claim to regress to previous lives when under hypnosis, in a trance, or meditating, few can back up such claims with concrete evidence. This disconcerting ability to speak an *unlearned* foreign language, known as xenoglossy (from the classical Greek for "strange tongue"), is very rare. In most cases, what was claimed to be an unlearned foreign language has proved to be mere gibberish or simple phrases learned long ago that seemed to bubble up from the subject's subconscious.

The 20th-century religious mystic Thérèse Neuman is reported to have relived the Passion of Christ by speaking in ancient Aramaic. Investigators claimed Neuman's was a genuine case of xenoglossy; others pointed out that all the phrases she had uttered appeared in print — in the Bible and elsewhere — along with translations.

On another occasion a 30-year-old French woman declared she was able to write Greek without ever having studied the language. It was later proved that all the phrases she recorded had appeared in a single Greek textbook to which she may have had access.

Unknown language

Yet xenoglossy should not be confused with glossolalia, or "speaking in tongues," the act of speaking in an unknown language which is most commonly associated with religious ecstasy. Various linguists who have investigated glossolalia report that these sounds, or "tongues," are not in fact true languages but merely language types, or as one linguist has noted, "refined gobbledygook."

Outside of hypnosis

Xenoglossy, however, has also been reliably reported occurring outside of hypnosis. In 1974 a 33-year-old Indian woman, Uttara Huddar, began speaking fluent Bengali while she was meditating during a stay in hospital. She called herself Sharada and could converse and write only in Bengali, a language allegedly unknown to her and her family.

How can xenoglossy be explained? Can it be that individuals like Dolores Jay and others are doing no more than reading the minds of the personalities they appear to be reproducing? Or are they merely generating elaborate feats of memory? Alternatively, despite strong evidence to the contrary, these might be extraordinarily sophisticated hoaxes. Whatever the answer, one thing is certain: further professional research needs to be directed at this mystifying, yet fascinating, phenomenon.

> **Religious mystic Thérèse Neuman is reported to have relived the Passion of Christ by speaking in ancient Aramaic.**

CRYPTOMNESIA

Many researchers believe that individuals who claim to speak foreign words or phrases they have never learned — or even heard — have merely forgotten that they learned them years ago. This unconscious memory of information that was learned in the past through normal channels is called cryptomnesia.

In the course of our normal lives we take in an awesome amount of information. If we did not consciously forget most of this information, other than that which is of some practical use or emotional significance to us, our minds would become hopelessly cluttered. The buried information may resurface when a person is under hypnosis, or in a trancelike state, and may then appear to him or her as totally new.

Minute details

Our hidden memories appear to be spectacular in both number and detail: under hypnosis, some subjects are able to bring back to mind minute details of books, newspapers, movies, and events they once witnessed — perhaps for only a second — years before. It is therefore not surprising that in some cases certain foreign words and phrases may surface again, unsummoned, from this vast store of buried memories.

ALIEN YET FAMILIAR

People have always seen strange objects in the skies. Not surprisingly, they have interpreted such phenomena in terms of their understanding of the world.

Mystery lights
In medieval times our ancestors interpreted strange signs and inexplicable objects in the sky as divine portents.

I N HIS *HISTORIA FRANCORUM* (*History of the Franks*), the French saint and bishop Gregory of Tours described how in A.D. 584 "there appeared in the sky brilliant rays of light which seemed to cross and collide with one another." The following year he recorded the appearance of "rays or domes...which seemed to race across the sky." For St. Gregory these lights were undoubtedly signs of some divine origin.

A little over two centuries later, the *Anglo-Saxon Chronicle* reported that in A.D. 793, flashes of lightning and fiery dragons were seen flying through the air in the north of England. These signs were interpreted as portents of

> "There appeared in the sky brilliant rays of light which seemed to cross and collide with one another."
> **St. Gregory of Tours**

divine displeasure. In A.D. 900, a flying ship was seen above the city of Lyon in France. For the people of the 10th century, this unidentified craft represented some demonic visitation.

The mystery airship

Between 1896 and 1897, a great airship — described by hundreds of witnesses as resembling a large cigar-shaped object that moved slowly through the night skies — appeared in the Midwest, Texas, and California. However, no such airship existed in the United States at that time.

Such sightings of otherworldly craft have continued right into the 20th century. In 1946, for example, Swedish authorities received nearly a thousand reports of mysterious rockets in their skies, some of which crashed into inland lakes. Swedish aviation engineers, who had considerable knowledge of the then-famous German V-2 rockets, failed to produce an explanation for these craft.

Visitor from Venus

By the 1950's extraterrestrial visitors were reported to be emerging from their spacecraft to make contact with earth-dwellers. At 12:30 P.M. on Thursday, November 20, 1952, in the California desert between Desert Center and Parker, Arizona, George Adamski claimed to be the first person to have made contact with an alien visitor. Adamski also claimed that he had communicated telepathically with this handsome alien visitor who had voyaged to earth from the planet Venus.

In 1980, Swiss farmer Willy Meier claimed that, between January 28, 1975, and March 4, 1978, he had had 105 encounters with extraterrestrials from the Pleiades star cluster, which is some 500 light years from earth. The alien visitors' spacecraft, Meier reported, were capable of traveling faster than the speed of light. What unites all these bizarre sightings of alien visitors is as simple as it is beguiling: If we place each of these sightings in its appropriate historical context we can also see there is a link between them.

Inexplicable objects

A sixth-century holy man, such as the French saint Gregory of Tours, would see a UFO as something essentially supernatural, while the pragmatic farmers of 19th-century America saw flying machines as yet another achievement of the golden age of terrestrial engineering. However, since the beginning of the space age, such mysterious lights and inexplicable objects have been identified habitually in the popular mind with extraterrestrial visitors. In each case, UFO sightings appear to represent those achievements that are just beyond humanity's own current capabilities.

> Since the beginning of the space age, such mysterious lights and inexplicable objects have habitually been identified in the popular mind with extraterrestrial visitors.

Fear or fantasy

Therefore the way in which we choose to interpret such bizarre phenomena in the skies tells us more perhaps about the development of civilization than it does about the possibility of the existence of spacecraft and visitors from other worlds. It may be that people are simply seeing natural phenomena, then coloring them with their wishes, fears, or fantasies. In this case, the explanation for this continuing enigma may, in fact, lie in the human mind, a mind that continually strives to overcome and outreach its limitations.

UFO's today
The human imagination still chooses to see alien craft in the skies even though modern scientific research provides little evidence for their existence.

CHANCE IN A MILLION

Each of us can relate an amazing coincidence that we have experienced. Some in the following collection are comic; some are tragic. The question still remains: If such events occur, can anything in our destinies be regarded as truly random?

IN 1984 A LOS ANGELES businessman, whom news reports identified only as Arnold G., was delighted to announce his engagement to his girlfriend, Carol. However, his fiancée's father immediately took Arnold to one side and confided that he was not the girl's natural father. Carol had been conceived by artificial insemination from donor sperm.

When Arnold discovered the name of the sperm bank, he was appalled: he had donated sperm there as a student. Arnold immediately obtained court authority to inspect the sperm bank's records. He discovered that he was the biological father of 807 children and that one of these was his fiancée, Carol. The dismayed couple's engagement was immediately broken off

Rare seabird

British naturalist Richard Watling went to Fiji in 1983 in the hope of tracking down an extremely rare seabird, MacGillivray's petrel. The small dark brown bird had been seen only once before, in 1855, when T. M. Rayner, the medical officer from the survey ship H.M.S. *Herald*, found a fledgling on the island of Gau in the Pacific. (The bird was named after the naturalist John MacGillivray, who was on the *Herald* at the time.)

Watling set up an elaborate system of lures — using flashlights and amplified recordings — to entice the petrel to come inland from the sea. After 12 months of constant, unsuccessful vigilance, he was beginning to think that the creature had become extinct. However, on the evening of May 9, 1984, a bird crashed into Watling's head in the dark. When he examined the creature, which was dazed by the collision, he identified it as a MacGillivray's petrel. Watling allowed the petrel to recover its equilibrium, and then let it go.

Icy irony

In December 1990, an international conference of the world's leading orthopedic surgeons took place at a clinic near the Swiss ski resort of Davos. On the second day of the conference, there was a sudden drop in temperature, and the paths around the clinic became iced over. Seven surgeons slipped on the ice and broke bones. The subject under discussion at the conference happened to be advanced bone fracture surgery.

Spinning fortune
Coincidences are startling because they involve the meaningful collision of apparently unconnected events or individuals. Such synchronicities prompt us to wonder if they are the result of forces that humankind does not yet understand.

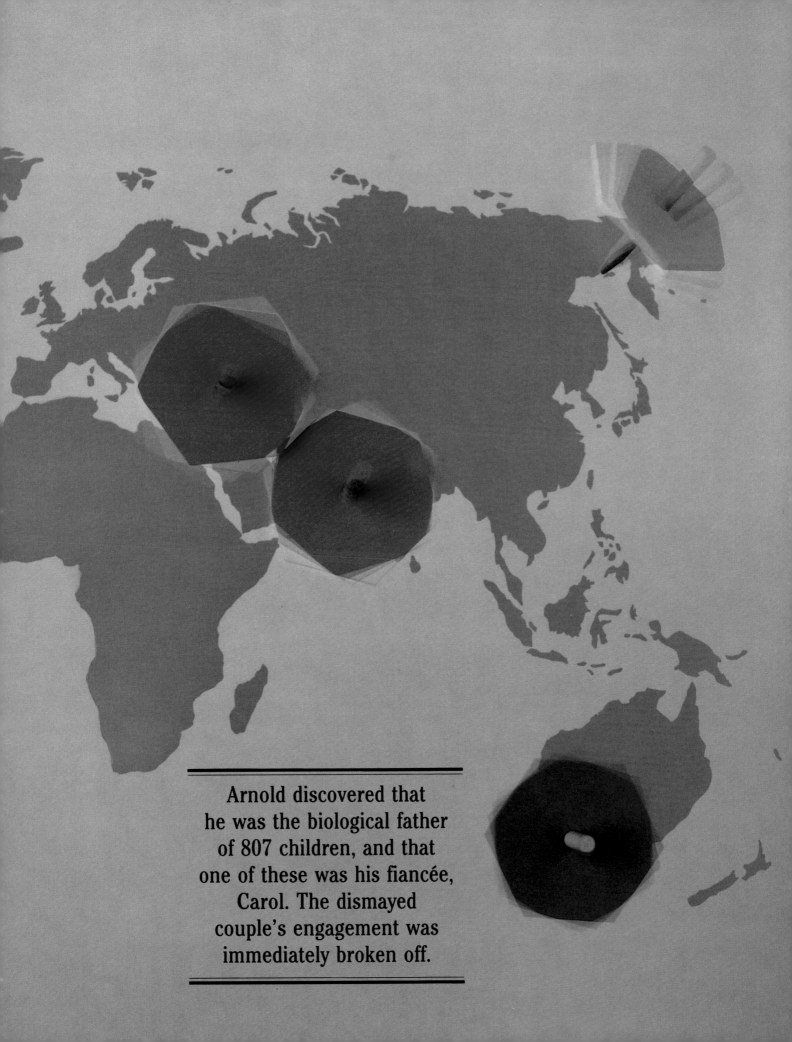

Arnold discovered that
he was the biological father
of 807 children, and that
one of these was his fiancée,
Carol. The dismayed
couple's engagement was
immediately broken off.

A long shot
The death of Gen. John Sedgwick at the hands of a distant sniper at Spotsylvania during the Wilderness Campaign of the American Civil War.

Red with embarrassment?
Members of the San Antonio Professional Firefighters Association cool off the parking lot after their cookoff. The event was to raise money for people who had lost their homes in fires.

One day in June 1988, school bus driver Lillie Baltrip was taking 29 other school bus drivers to a ceremony in Houston, Texas, at which she was to receive an award for an unbroken record of safe driving. However, on this occasion Lillie took a corner too sharply and tipped the bus over. Lillie and 16 of her passengers were injured and briefly admitted to hospital. Fortunately, neither Lillie nor any of her passengers sustained serious injuries.

Fire! Fire!

Firefighters have also been the victims of embarrassing coincidences. In June 1974, firefighters at the Black Sea resort of Sukhumi, in the Soviet republic of Georgia, thought three calls telling them that their station was on fire were false alarms. Unconcerned, the firefighters continued playing dominoes while the fire station burned down around them.

In March 1986, the local paper in Lingenfeld, Germany, carried this notice: "The information evening planned for Saturday in the Weingarten fire station on 'Preventative Measures for Fighting Fire in House and Home' has had to be postponed. The fire station caught fire on Monday and must first be restored."

On March 3, 1991, the San Antonio Professional Firefighters Association organized a feast — the Third Annual Bush's Canned Foods South Texas Cookoff — to raise money for people whose homes had been destroyed by fire. Eight thousand people turned up to partake of the feast. However, in the middle of the party, the dry grass in the parking lot caught fire. The ensuing conflagration destroyed 97 vehicles.

> **English novelist Arnold Bennett foolishly drank a glass of water in Paris to prove how safe it was. He caught a fatal dose of typhoid.**

Sometimes fate delivers, metaphorically speaking at least, a kick in the teeth. In May 1864, during the Civil War Battle of Spotsylvania, Virginia, Union general "Uncle John" Sedgwick looked across at the entrenched forces of Confederate general Robert E. Lee, and commented:

"They couldn't hit an elephant at this dist...." Before he could complete his sentence, Sedgwick fell dead from an enemy sniper's bullet. In another twist of fate, the English novelist, Arnold Bennett, foolishly drank a glass of water in Paris, in 1931, to prove how safe it was. He caught a fatal dose of typhoid.

Alpine storm

More recently, on July 1, 1985, Helena Burgler of Illgau, Switzerland, went to the village's Sacred Heart Chapel to warn the local people of an approaching electrical storm. (There was an old tradition in this Alpine village to warn fellow villagers of impending bad weather.) Once Helena, who was the mother of 13 children, had delivered her message, so the story goes, she stepped out of the chapel and was immediately struck dead by a bolt of lightning.

Other people have tempted fate by being too clever for their own good. In 1986 Ed Morris of Seattle went moose hunting with some friends in the woods of northern Canada. Intending to amuse his friends, he slipped away and put on a novelty hat with foam rubber antlers. Then, hiding among the trees, he began moaning like a moose. Instead of laughs, he got buttocks full of buckshot.

Tragic camouflage

The consequences of adopting camouflage were more serious for 43-year-old Charles Boyer when he went turkey hunting in Deerfield Township, Pennsylvania, in November 1990. Boyer daubed his face and clothing with blue and gray patches of paint to look like a turkey. Hoping his appearance would enable him to get closer to the birds, he crouched behind some bushes and made gobbling noises like a turkey.

Unfortunately, Boyer's plan did not work quite as he had anticipated. Fellow hunter Troy Moore, from a distance of 90 yards, spotted what he thought was a turkey, took aim, and fired. His shot killed the unfortunate Charles Boyer.

The Cosmic Banana Skin

We can, of course, interpret all these incidents as reflecting the randomness and cruelty of fate. Moreover, if we follow Charles Fort's interpretation of the

LOST AND FOUND

Against all the odds, some lost or discarded articles find their way back to their former owners, almost as though they possess the bizarre ability to hunt them down.

LEGEND HAS IT THAT, in the seventh century A.D., the Scottish king Cadzow, who ruled the kingdom that is known today as Strathclyde, began to suspect that his wife was having an affair with a knight at his court. In happier times, the king had given his wife a gold ring. Now the queen no longer wore the ring and Cadzow began to think that she had given it to her new lover. The king sought out the knight he suspected; he found him sleeping by the river Clyde. Taking care not to wake him, the king searched the knight, and found the ring. In a rage, he hurled it into the river.

Returning to Glasgow, his capital city, the king sent for his wife. "This evening, wear the gold ring I gave you," he requested. Now at the end of her tether, the queen hastened to Mungo, who was the bishop of Glasgow, and threw herself upon his mercy. Mungo's name means "dear one" and the queen's trust in this good man was not misplaced.

Mungo sent a monk to fish in the river, the legend says, instructing him to bring back his first catch, which was a salmon. When Mungo opened the fish's mouth he found the queen's ring. Mungo was later canonized and became the patron saint of Glasgow. Today two salmon with rings in their mouths are incorporated in the city's coat of arms in memory of this legend.

Miraculous returns

This is merely a legend, but there are equally amazing reports of things lost and found again today. One day in 1972, Ricky Shipman lost his wallet from a pocket in his swimsuit while swimming with friends off a beach in North Carolina. Eleven years

Fishy story
The legend of St. Mungo is commemorated in the coat of arms of the city of Glasgow.

later a man named Gause, the owner of a restaurant in Little River, just across the state line in South Carolina, returned Shipman's driving license, which had been in his lost wallet. A friend of Gause had caught a large Spanish mackerel near the beach where Shipman had been swimming, and, gutting it, he had found the license inside, protected by its plastic cover.

Ring back

In 1961, while on holiday in St. Anne's, Lancashire, England, Brenda Rawson, who was engaged to Christopher Firth, lost the diamond ring he had given her. In 1979 she was having a casual conversation with her husband's cousin John about metal detectors when John remarked that 18 years earlier one of his children had found a diamond ring at St. Anne's. It turned out to be Brenda's.

In 1982 another lost ring made the most timely of reappearances. Albert Thornton just happened to be gardening at his home in Tolworth, Surrey, England, when he dug up his wife's wedding ring, lost 15 years earlier. He found it, so the story goes, on the couple's silver wedding anniversary.

Marine eyeglasses

In September 1988, Gosselin Delius of Brussels, Belgium, was sailing a yacht through a Force 7 gale off the coast of Folkestone, in the south of England, when in the storm he lost his eyeglasses overboard. Some weeks later he happened to read in a newspaper that Belgian fisherman Yan Gazelle had caught a 13-pound monkfish, gutted it, and found a pair of eyeglasses inside. Delius, pursuing what must have seemed like a long shot, contacted the fisherman and found that the spectacles bore the same serial number as his own. They were slightly bent but still usable.

Sham marriage
Peter and Malika Reyn-Bardt on their wedding day in 1959. Theirs was a marriage of convenience.

universe — the idea that such events are the work of the Cosmic Joker, that transcendent trickster who toys with the fate of each of us — we can detect a bit of macabre humor. If we can overlook the tragic death of others, in the great cosmic scheme of things, such ironies, from the Fortean viewpoint, may serve a broader purpose. They remind us of the possibility that there exists some larger force at work in the universe, a force greater than our own selfish interests.

Avenging fate
Yet other extraordinary coincidences suggest that whatever it may be that orders our lives, this force perhaps does not overlook injustice or wrongdoing.

On May 13, 1983, a man cutting peat (used as fuel for fires and as humus in which to grow plants) in the Lindow Moss peat bog in Cheshire, England, found a skull. One eye was still intact and hair was still attached to the grisly relic. That such human remains should be found in the Lindow Moss peat bog was not without precedent. Complete bodies have been known to be perfectly preserved for thousands of years in the peat bogs of Denmark and Ireland.

However, initial forensic tests on the skull seemed to indicate that it had been buried for at least 5, but no more than 50

> # A man cutting peat found a skull. One eye was still intact and hair was still attached to the grisly relic.

years. The partly preserved head was that of a European woman aged between 30 and 50. Local police checked their missing persons files, and their attention was caught by the disappearance of Malika Reyn-Bardt, wife of Peter Reyn-Bardt, in 1961. The skull had been found near the cottage where former airline pilot Reyn-Bardt had lived 20 years ago.

Final separation
In June 1983, police interviewed Reyn-Bardt. He gave a clear account of the last time he had seen his wife. Reyn-Bardt said he had married her in 1959 but only to gain respectability with the airline for which he worked, because he was a homosexual. The couple had separated by the end of 1959. The Reyn-Bardts had not seen each other until one day in June 1961, when Malika arrived at the cottage near Lindow Moss. According to Reyn-Bardt, his wife had threatened to expose him as a homosexual to his employers unless he gave her money. He sent her away with a 10-pound note.

The police confronted Reyn-Bardt with the skull from Lindow Moss. He changed his story, confessing that he had actually strangled Malika, cut her body up, and buried the pieces near his house. That was good enough for the purposes of the police, and Reyn-Bardt was charged with his wife's murder.

The preserved head was then dispatched to Oxford University's archeological research laboratory for further tests to establish the date of the skull. The results, announced on October 12, 1983, revealed beyond any doubt that the skull was far older than the first analysis had suggested: It was the skull of a European woman who had died in about A.D. 410, at the time when the Romans were leaving the area.

Reyn-Bardt pleaded not guilty, but was convicted of murdering his wife and sentenced to life imprisonment at his trial the following December. This was despite the fact that no trace has ever been found of Malika Reyn-Bardt's body. Friday, May 13, 1983, was certainly not a very lucky day for Peter Reyn-Bardt.

ANIMAL REVENGE

Hunted animals, dead ones as well as live, have often been known to injure, or even to kill, their human predators. Acts of animal poetic justice — perhaps?

EARLY ONE EVENING in December 1987 hunter Ray Canny, rifle at the ready, crouched in a ditch beside a field near Osage, Iowa. His sons were flushing deer from the other side of the field and Canny was lining them up in his sights. Unfortunately for the hunter on this occasion, one of the animals his sons was pursuing attempted to leap the ditch Canny was hiding in and landed on him, breaking his neck.

Though very rare, such bizarre incidents are construed by some extreme opponents of hunting, shooting, and fishing as justified animal revenge on those who set out to kill them for amusement. The following are chance-in-a-million examples of animals turning the tables on their tormentors.

Knockout game

In England, in the same month that Canny met with his accident, a dead grouse took its revenge on its killer. Gilbert Fenwick was shooting on Lord Bolton's estate at Wensley in the Yorkshire Dales. Having shot one bird, he was taking careful aim at another when the first, weighing a pound and a half, hit him in the face at an estimated speed of 60 m.p.h. and knocked him unconscious.

When this incident was reported to The British Field Sports Society, they could find no precedent for Gilbert Fenwick's accident. However, a few days later the British author Chapman Pincher wrote to the London newspaper the *Daily Telegraph* to say he had suffered a similar mishap in 1961 while shooting on Sir Thomas Sopwith's estate at Arkengarthdale in Yorkshire: a dead grouse had fallen on him and broken his nose.

Wahoo yahoo

On March 1, 1987, Lou Wiezai was fishing from a boat in the Pacific, about 250 miles south of Baja California, Mexico, when a wahoo — a large game fish of the mackerel family — leaped out of the water and bit into his left hand and forearm. Its razor-sharp teeth cut the flesh to the bone, and Wiezai had to be rushed to a hospital for emergency treatment to his wound.

Some creatures have even appeared to collaborate against their human tormentors. When fishing in the River Idle, Nottinghamshire, England, in September 1988, British angler Dave Martin disturbed a wasps' nest hidden in the rotten log he was sitting on. The wasps retaliated, finding their way inside Martin's clothes. He leaped into the river, but he found he was not welcome among the fish he had been hoping to catch: A large pike bit through his pants and slashed his leg.

SON OF SAM

From 1976 to 1977 a psychopath calling himself the Son of Sam committed an appalling series of murders in the city of New York. As the case unraveled, it began to display a macabre, and exquisitely bizarre, set of coincidences.

IN JULY 29, 1976, at about 1 A.M., Donna Lauria, an 18-year-old student, was shot and killed instantly as she sat in a parked car talking through the open window to a friend, Jody Valente. Jody, who was standing outside and was shot in the leg, described the killer to the police as a white male, in his 30's, stocky, husky-voiced, and very calm. He had walked, not run, from the scene of the murder.

Random killings

Twelve more shootings, resulting in five more fatalities, took place in the next 12 months. Finally, on August 10, 1977, 24-year-old David Richard Berkowitz was arrested and charged with the murders. The randomness of the killings had appalled and caused terror among New Yorkers for over a year. That evening the ABC TV network devoted half of its nationwide news broadcast to reporting the details of the final capture of the Son of Sam, as the killer called himself.

The police investigation into the Son of Sam case turned up many sinister synchronicities. But as more became apparent about the killer, these became less attributable to the workings of the Cosmic Joker than to the macabre obsessions of a seriously twisted mind.

New York City street map

Arrest of the Son of Sam
David Berkowitz (center) was imprisoned for 315 years for the murders.

DOG CONNECTIONS

◆ Berkowitz claimed he received his orders to kill from a dog called Sam. This dog was owned by Sam Carr, Berkowitz's neighbor in the New York City district of Yonkers.

◆ On December 24, 1976, Berkowitz shot a dog named Rocket at 18 Wicker Street, Yonkers. Two days later, three German Shepherd dogs were found slain in a gutter near Wicker Street.

◆ Early in 1977 Berkowitz applied for a position at a Yonkers dog kennel.

◆ Donna Lauria's father, Mike, was bringing her pet poodle down from the apartment as his daughter was shot. All the murders were committed with a Bulldog .44 magnum revolver.

◆ While Berkowitz was in police custody, three more German Shepherd dogs were killed in one incident in Yonkers.

THE WICKER MAN CONNECTIONS

◆ In 1973 a movie called *The Wicker Man* was released. It was set on a remote island in Scotland and starred Britt Ekland and Edward Woodward. It told the story of the ritual murder of a virgin policeman by the pagan villagers. He meets his gruesome end when he is burned alive in an enormous wicker effigy.

◆ In April 1977, during the height of the Son of Sam murders, *The Wicker Man* was rerun at a Manhattan movie theater.

◆ A few weeks later, in May 1977, Berkowitz firebombed the house of Joachim Neto, the owner of Rocket (the dog Berkowitz had shot in December 1976). Neto lived at 18 Wicker Street.

◆ On June 5, 1977, a letter was sent to the New York *Daily News* columnist Jimmy Breslin from someone who had inside knowledge of the murders. The letter, portions of which were published, ended with occult hieroglyphics and was signed "Son of Sam, The Wicked King of Wicker."

◆ Berkowitz lived in an apartment at 35 Pine Street in Yonkers. His street was only one block away from Wicker Street.

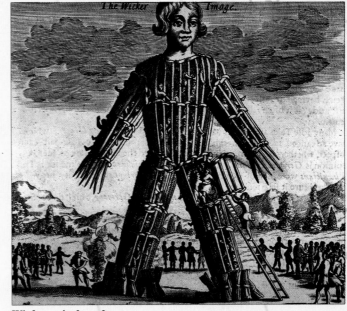

Wicker witchcraft
This illustration, from Sammes' Britannica *(1676), shows the pagan ritual of seasonal human sacrifice.*

The demon Behemoth

OCCULT CONNECTIONS

◆ In a letter to the police found at the scene of the Son of Sam murders of Valentina Suriani and Alexander Esau on April 17, 1977, the killer ended with the words: "I am the 'monster' – 'Beelzebub' the 'Chubby Behemoth.'"

◆ At the bottom of the letter sent to journalist Jimmy Breslin was drawn an occult symbol, or sigil, created by the 19th-century French occultist, Alphonse-Louis Constant, who called himself Eliphas Levi. "Eliphas" was a play on the Latin word for elephant, *elephas*. The demon associated with the elephant in the occult tradition is "Behemoth."

◆ On June 25, 1977, Judy Placido and Salvatore Lupo were shot in the parking lot after a night's dancing at the Elephas Disco in the Bayside section of Queens.

◆ The drawings of Eliphas Levi include an elaborate design for the evocation of demons. The demons' names include "Berkaial" (Berkowitz?) and "Amasarac." The latter name can be interpreted as an approximate reverse spelling of Sam Carr, the name of the owner of the dog that Berkowitz claimed gave him orders to kill.

FORTUNE OR FATE?

Could meaningful coincidences be manifestations of some unfathomable linking principle in the universe, one that owes nothing to the system of cause and effect on which the physical laws operate that govern our everyday reality?

YOU ARE READING IN A BOOK an uncommonly used word, such as *heliotrope*, when, at the selfsame moment, you hear that word spoken, in conversation or on the television or radio. Such insignificant coincidences are familiar to all of us, but we shrug them off as attributable purely to chance.

Golden scarab

One such coincidence occurred while the celebrated Swiss psychologist Carl Gustav Jung was seeing one of his patients. As he described in his book *Synchronicity: An Acausal Connecting Principle* (1952):

> "A young woman I was treating had, at a critical moment, a dream in which she was given a golden scarab. While she was telling me this dream I sat with my back to the closed window. Suddenly I heard a noise behind me, like a gentle tapping. I turned round and saw a flying insect knocking against the window pane from outside. I opened the window and caught the creature.... It was the nearest analogy to a golden scarab that one finds in our latitudes...which contrary to its usual habits had evidently felt an urge to get into a dark room at this particular moment."

This incident convinced Jung that some greater reality existed that linked these coincidences together. Jung explained coincidences thus: "The more they multiply and the greater and more exact the correspondence is, the more their probability sinks and their unthinkability increases, until they can no longer be regarded as pure chance but, for lack of a causal explanation, have to be thought of as meaningful arrangements."

When Jung developed his theory of the collective unconscious...he probed beyond Western civilization, and examined other cultures to develop his ideas. He was drawn to astrology and... to the *I Ching*, the ancient Chinese book of divination.

Chinese book of divination

When Jung developed his theory of the collective unconscious (the idea that certain images that he called "archetypes" are common in all human thinking) he probed beyond Western civilization, and examined other cultures to develop his ideas. He was drawn to astrology and, in particular, to the *I Ching*, the ancient Chinese book of divination.

This method of divining involves the creation of one of 64 hexagrams through the casting of yarrow stalks or the throwing of coins. (The fall of the sticks, or the coins, leads to the formation of a six-line figure, or hexagram.) Each hexagram has its own detailed, line by line, commentary, which, when interpreted, is believed to give the questioner clear guidance as to how to proceed in relation to the question asked.

For Jung it was clear that when the ancient Chinese consulted the *I Ching*, they accepted the chance aspect of events. Following his theory of synchronicity, Jung believed that, in that very moment when a hexagram was formed, a relationship existed between the answer to be found in the hexagram and real events. Jung further argued that when the questioner consulted the *I Ching*, the meaning in the question would also draw it to the most appropriate of the 64 possible hexagrams.

Difficult concept

We find this unfamiliar concept difficult to understand from our 20th-century viewpoint, because we are more likely to interpret chance events as mere coincidences. Jung argued, in turn, that our Western world's ingrained belief in cause and effect made it difficult for us to accept that causeless events might ever happen. A question Jung's theories should prompt us to consider is this: Why, when unexpected events happen so frequently, have we not learned to expect them?

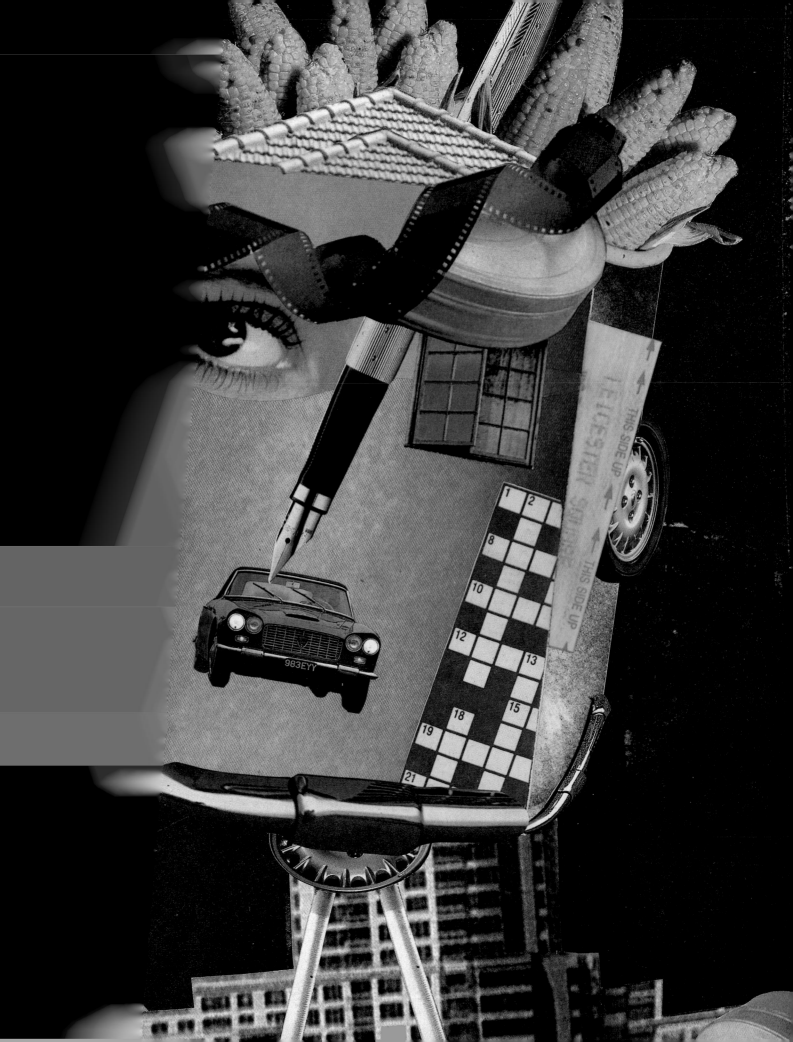

CHAPTER FOUR

EYE OF THE BEHOLDER

Why, when you buy a car, do you suddenly start seeing that same make of vehicle everywhere on the streets? We all have observed similar things in our daily lives. Clearly, our viewpoint can lead us to make connections between seemingly unrelated phenomena, and thus affect the way we experience our world.

In 1971 the Oscar-winning British actor Anthony Hopkins auditioned for the leading role in *The Girl from Petrovka*, a film that was based on the book of the same name by George Feifer. Hopkins was chosen for the part and a few days after he had signed the contract, he traveled to London hoping to buy a copy of the book. Having failed to find one after visiting several secondhand bookstores in the

COINCIDENCES AND SYNCHRONICITY

Many people's lives are marked by the phenomenon of coincidence. In 1919 the Austrian biologist Dr. Paul Kammerer published *Das Gesetz der Serie* (*The Law of Seriality*) in which he attempted to define the laws of coincidence. Kammerer argued that coincidence is the manifestation of a natural principle that operates independently of any known physical causation.

These ideas were further explored by the Swiss psychologist Carl Gustav Jung, who in 1952 coined the term *synchronicity*, which he defined as "the simultaneous occurrence of two meaningful but not causally connected events....a coincidence in time of two or more causally unrelated events which have the same or similar meaning."

Lightning strikes

On June 6, 1969, the day Jung died, a great thunderstorm raged around his house at Kusnacht in Switzerland. Jung's biographer, Laurens van der Post, recorded that "Just about the time of his death, lightning struck his favourite tree in the garden." Following Jung's theory of synchronicity, we might interpret the connection between the natural forces and his death as meaningful in that it suggests how the human psyche is interrelated with the external world. Jung's theory of synchronicity, however, still remains unproven.

Charing Cross area of London, he was waiting on the platform of Leicester Square subway station when he noticed a book had been left on the seat next to him. He picked it up and discovered it was a copy of *The Girl from Petrovka*.

Two years later, while in the middle of filming in Vienna, Hopkins was visited on the set by Feifer. The author mentioned that he no longer had a copy of his own book. Feifer told Hopkins how he had lent his last copy to a friend who had mislaid it somewhere in London. Then Feifer added that it had been especially annoying because he had annotated that particular copy.

Scarcely believing that such a coincidence could be possible, an astonished Hopkins handed Feifer the copy he had found at Leicester Square subway station. "Is this the one?" he asked, "with the notes scribbled in the margins?" It was indeed the very book Feifer's friend had lost.

Celluloid connections

The British actress Julie Christie is another Oscar-winner for whom life appears, on one occasion, to have

Heart-rending moment
In this scene from Don't Look Now, *Donald Sutherland plays the father who carries his drowned child from the pond. A similar real-life tragedy later touched the life of his costar in this movie, Julie Christie.*

imitated art. In 1973 Julie Christie starred in the compelling movie *Don't Look Now* that was dramatized from a story by Daphne du Maurier. In this movie Julie Christie and Donald Sutherland played the parents of a child who is tragically drowned in a shallow pond in the garden of their country house.

Six years later Jonathan and Lesley Heale were renting their friend Julie Christie's farmhouse in the Welsh countryside. The movie star had just left after a visit when Lesley Heale suddenly noticed that her 22-month-old son was missing. She found him dead, as in the movie, floating face down in the duckpond in front of the house.

All in a name

Another less than happy but curious sequence of events occurred on August 3, 1979, in the town of Riom, France. A car driven by Jean-Pierre Serre of Paris crashed into a car driven by Georges Serre of Clermont-Ferrand. Seconds later, a car driven by Mlle. C. Serre of Royat

Anthony Hopkins

crashed into the other two. Police investigations showed the three drivers were not related and had never met before. Moreover, the surname Serre is not particularly common in France. As a matter of fact, there are only about 120 Serres listed among the 1,250,000 entries in the Paris telephone directory.

Food for thought

For some people, however, it appears that a name can have a detrimental effect on their behavior. In February 1979 an American Indian was indicted in Texas for the attempted theft of a pack of cigarettes. His name was Stanley Joseph Stillsmoking. In September 1988 an undertaker from Universal City, California, caused a disturbance on a flight from Boston to Los Angeles, when a stewardess told him his seat had been redesignated as non-smoking, and that he must therefore extinguish his cigarette. His name was James Tobacca.

Inanimate objects can also be the subject of weird coincidences. In June 1979 a truck loaded with 40 tons of carrots crashed into a truck carrying an equal quantity of salad oil on Leek Road near the Santa Ana Freeway outside Los Angeles. "The biggest carrot salad ever," was how a bemused highway patrolman described the vast vegetable pile. A month later another culinary collision occurred when a truckload of broiling chickens smashed into a tanker carrying barbecue sauce on the George Washington Bridge in New York City. The ensuing conflagration resulted in a gigantic serving of flame-cooked barbecue chickens.

In September 1982, near the village of Stoneycross, in the south of England, a truck overturned, spilling a ton and a half of live eels. It just so happened that the truck had collided with a car bearing the license plate EEL 293V.

Hidden treasure

Most people, however, would be happy to experience what happened to a farmer from Dorset in the southwest of England. On May 22, 1983, Simon Drake had just finished plowing a field

> **An American Indian was indicted in Texas for the attempted theft of a pack of cigarettes. His name was Stanley Joseph Stillsmoking.**

REVERSE HILLS

There are places where gravity, or some less familiar force, seems to pull things uphill. At Croy Brae in Strathclyde, Scotland, the road going north out of town looks as if it slopes uphill. Yet if a driver turns off the engine and freewheels, the car picks up speed as it rolls toward what appears to be the brow of the hill. At Magnetic Hill, near Moncton, New Brunswick, Canada, drivers go down what looks to be a shallow slope into a dip in the road. But if drivers change into neutral gear, they find their cars roll backward, apparently uphill, at speeds of up to 15 m.p.h.

Optical illusions

Scientific investigations have ruled out the action of magnetic forces on the vehicle as the cause of this phenomenon. The solution to these "reverse" hills, it appears, must be sought in the perception of the individual involved. It is well known how easily the human eye may be deceived by optical clues. A road that actually descends can appear to do the opposite if the surrounding terrain is sloping steeply downhill. This phenomenon is familiar to golfers trying to judge the slope, or "borrow," when putting on greens on hilly courses.

Local magnetic anomalies might also provide an explanation for such phenomena. They might interfere in some way with the balance mechanisms in the ear.

Treasure trove
At auction in 1985, each of the gold coins found by Simon Drake fetched over $1000.

DiBiasi of Hines, Oregon, who owed the taxes. The confusion arose because both women were born Patricia Ann Campbell, on March 13, 1941, and they also shared the same Social Security number. Both women's fathers were named Robert; both women married men in the military within 11 days of each other in 1959. If that was not enough of a coincidence, both Patricias also worked as bookkeepers and had children aged 21 and 19. Investigations revealed that Patricia Kern and Patricia DiBiasi were not related and had never met.

Eerie repetitions

Camille Flammarion, the French writer and astronomer, records the following tale in his book *The Unknown* (1900). He was sitting at his desk working on a chapter about the force of the wind for

Camille Flammarion
This 19th-century French astronomer was a prodigious writer in his field. He was also a student of the occult and, in later life, became interested in psychical research.

on his farm when he kicked the soil to see if it was ready for sowing his spring barley. This was a variety of barley called Golden Promise. Suddenly, as he looked down, he noticed a bright metal disc in a newly plowed furrow. He took it up to the farmhouse and looked it up in a book on coins. He realized that he had found a gold noble, dating from the 15th century.

Using a metal detector, over the next couple of months Drake found a total of 100 coins in the same field. After an inquiry, the coins were declared treasure trove and were returned in lieu of a reward to their finder. In May 1985, Drake sold 95 of the coins at auction at Christie's in London for $100,000. Golden Promise indeed.

The taxman cometh

An error by the U.S. Internal Revenue Service (IRS) gave Patricia Kern of Aurora, Colorado, a shock. In 1983 the IRS sent her a tax request for $3,000 for a job in Oregon, a state she had never set foot in. Inquiries showed that the demand was intended for Patricia

> **Suddenly Simon Drake noticed a bright metal disc in a newly plowed furrow. He looked it up in a book on coins and discovered that he had found a gold noble, dating from the 15th century.**

another book, *The Atmosphere* (1872). All of a sudden a "miniature whirlwind" whisked his sheaf of papers out of the window. Flammarion faced the worst disaster that can ever befall a writer — the loss of his manuscript.

A few days later Flammarion received the proofs for his book. Miraculously enough, they included the missing chapter. The gust of wind, it appears, had deposited the sheets of paper at the feet of the publisher's messenger, who just happened to be passing by at that very moment. The messenger gathered up the papers under the misapprehension that he had dropped them, and delivered them to the office where they were added to the rest of the manuscript.

OPERATION CROSSWORD

On June 2, 1944, the Allied Intelligence Chiefs feared that top secret plans for the most important military operation of the Second World War had been betrayed. Or was it just a startling set of coincidences?

ON MAY 3, 1944, the crossword puzzle in a London newspaper, the *Daily Telegraph*, carried the following clue for 17 across: "One of the U.S." (four letters). On May 23 the clue for three down read: "Red Indian on the Missouri" (five letters). On May 31 the clue for 11 across read: "This bush is a centre of nursery revolution" (eight letters). The solutions to these clues were, in the same sequence, UTAH, OMAHA, and MULBERRY. When an unnamed member of military intelligence in London who regularly completed the crossword puzzle saw these words, he at once drew them to the attention of his superiors.

On June 2, 1944, the crossword puzzle contained two further clues: Thirteen down read "Britannia and he hold to the same thing" (seven letters), and 11 across read: "But some big-wig like this has stolen some of it at times" (eight letters). The Allied intelligence chiefs were appalled. The solutions to these clues — NEPTUNE and OVERLORD — completed a horrifying revelation.

The crossword puzzle had, over a period of 31 days, revealed five of the top secret code words for the D-Day landings that were due to take place on June 6, 1944.

At this point the full might of military intelligence swung into action. Members of MI5, a branch of the British Secret Service, swooped down on the newspaper's offices, expecting to unmask a spy who was busy sending messages back to his masters in Berlin. The crossword puzzle, they found, had been composed by one Leonard Sidney Dawe, a teacher from Leatherhead, Surrey. He was the senior crossword compiler for the newspaper and had been composing crosswords for the *Daily Telegraph* for the previous 20 years. He was interrogated intensively by MI5, but in the end they had to conclude that Dawe was innocent of any espionage. It had all been a startling coincidence.

Allied success

The D-Day operation was code-named Overlord; two of the landing beaches in northern France were known as Utah and Omaha; while the artificial landing harbors were called Mulberry. The naval operation was code-named Neptune. During the military planning for the invasion, the operation and the landing sites had been referred to only by their code names.

Except for this short-lived crisis, the D-Day operation was one of the best kept secrets of the war, and the Allies' success on D-Day marked the beginning of the end of the Second World War in Europe.

THIS IS THE YEAR !

IT'S UP TO US TO LET 'EM HAVE IT !

D-Day landing
In the early hours of June 6, 1944, the day after the worst weather conditions in the English Channel for 25 years, the skies cleared and the Allied invasion of northern Europe began. By nightfall on the first day of Operation Overlord, the Allied forces had gained control of all five landing beaches.

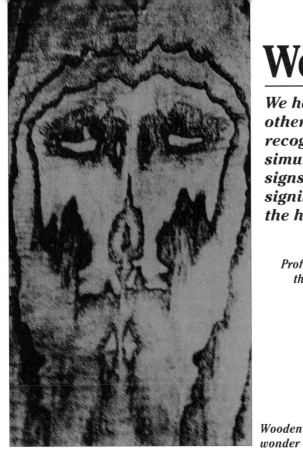

WONDERS IN NATURE

We have all seen bizarre shapes in inkblots, clouds, or other natural formations that appear to mimic recognizable forms. Such phenomena are known as simulacra. Humankind has often interpreted these signs as some miraculous manifestation. Yet, perhaps significantly, it is always how they are interpreted by the human eye that gives them form and meaning.

Profile in the sky

Wooden wonder

The Bacup plank
In November 1988 a workman at a factory in Bacup, Lancashire, in the north of England, saw this likeness of Christ in a plank of wood as he was about to cut it up. The face is reminiscent of the one found on the Turin Shroud. Moreover, on November 18, 1988, the day the Bacup plank was found, scientists announced that the Turin Shroud could not possibly be authentic.

Cloud person
This photograph, in common with many others of its kind, was taken to record a cloud formation. Only afterward, when the film was processed, was the face detected in the bottom center of the picture.

The Bear of Brimham
This simulacrum of a dancing bear with its paws in the air is a natural rock formation in Brimham, Yorkshire, in the north of England. Madame Helena Blavatsky, the founder of the mystical Theosophical Society, believed Brimham was the site of animal worship in ancient times.

Dancing bear

Forest spirit

What at first might appear to be the reflection on water of the vegetation of an Amazonian rain forest is, in fact, the manifestation of a divine message, according to an interpretation made by a shaman, or holy man, of the Iquito Peruvian Indians. Rotate this photograph clockwise through 90 degrees and the image becomes that of a monkey with tiny eyes. It is said to symbolize a man unable to open his eyes wide enough to understand the world around him.

Divine reflection

Legendary rock

Rock formations often seem to take on human forms, and sometimes these images are taken to be representations of local heroes or mythological figures. King Arthur, the legendary ruler of Britain, was said to have been born at Tintagel Castle in Cornwall, in the southwest of England. This sheer cliff nearby is known as King Arthur's Head.

Miraculous face

Stone martyr

Inserted into the stonework facade of the tower of St. Salvator's College, St. Andrew's, Scotland, is a brick that bears the distorted image of a face. Local legend relates that the face miraculously formed on the brick is that of the Protestant martyr Patrick Hamilton, who was burned at the stake in 1528.

Regal features

THE FAIRY FOLK

One summer afternoon during the First World War two girls claimed to have seen some fairies. Nobody would have believed them, of course, if they had not produced several photographs as proof!

FRANCES GRIFFITHS, AGE 10, and her cousin Elsie Wright, age 16, lived in the Yorkshire village of Cottingley in the north of England. In the summer of 1917 the cousins repeatedly told Elsie's parents that they had seen fairies down at the beck, the fast-flowing little stream that ran through the village. Frances regularly returned home with her shoes and clothing soaked from falling into the water there while, she claimed, she was watching the fairies.

The first photograph

The Wrights did not believe her story, but Elsie's father was persuaded to lend the cousins his camera the next time they went to the beck. He owned a box-type camera that could be loaded with a single photographic plate. On a hot Saturday in July 1917, the cousins spent the afternoon down by the stream. The girls took a photograph, which Elsie's father, Arthur Wright, developed later that day. Wright was puzzled to see certain white shapes taking form in the foreground. "Oh, Frances, the fairies are on the plate!" declared an excited Elsie. The photograph showed Frances's solemn face resting on her hands, and in front of her a ring formed by four dancing fairies.

Later that evening, Mr. and Mrs. Wright went through the attic bedroom the girls shared to see if they could find any evidence of the manufacture of fairylike figures. They had no success. Next they searched the area around the beck, but they could come up with no evidence that the photograph had been faked.

A dancing elf

Two months later Frances used the same camera and photographed Elsie sitting on the grass with her hand extended toward a little dancing gnome. Arthur Wright questioned the girls closely, but they did not change their story: they said consistently that they had simply photographed the fairies, elves, and gnomes they often saw at the Cottingley beck.

Fairies, it appeared, were a normal part of the girls' lives. On November 9, 1918, Frances wrote in a letter to a young friend: "I am sending two photos, both of me, one of me in a bathing costume in our back yard, Uncle Arthur took that, while the other one is me with some fairies down at the beck, Elsie took that one." On the back of the fairy photograph Frances, who had been born and lived in South Africa during her early childhood, had scribbled in a girlish scrawl: "Elsie and I

"One evening after I came home from school I went up to the beck to a favourite place — the willow overhanging the stream. Then a willow leaf started shaking violently — just one....As I watched, a small man, dressed all in green, stood on the branch with the stem of the leaf in his hand....He looked straight at me and disappeared."

Frances Griffiths

Fairies at rest
This enchanting watercolor of fairies asleep in the moonlight was painted by Charles Doyle in 1875 to illustrate William Allingham's poem In Fairy Land.

101

Leaping fairy
Elsie took this photograph of Frances watching a fairy in flight in August 1920. Elsie later revealed that she had actually drawn this fairy freehand, and attached it to the branch of a tree with a hatpin.

Truth about the fairies
In 1982 Elsie Wright, age 81, revealed how she had faked the Cottingley photographs. Here she shows what her cutout fairy drawing would have actually looked like.

are friendly with the beck fairies. It is funny I never used to see them in Africa. It must be far too hot for them there."

Elsie Wright's mother, Polly, believed in fairies, having long been interested in occult phenomena. In the summer of 1919 she attended a lecture on fairy life given at the Theosophical Society in Bradford. Founded by the Russian mystic Madame Helena Blavatsky, the Theosophical Society believed that thought forms and spirit worlds could be materialized, and that clairvoyants could see them.

An expert on fairy lore
At the lecture Polly spoke of the fairy photographs, and before long news of the pictures reached Edward L. Gardner, a leading Theosophist and expert on fairy lore. He contacted the Wrights at once, and on seeing the original photographs, he set about improving the quality of the prints. He was convinced that the pictures were genuine even though he never actually examined the original photographic plates.

At this point an eminent champion took up the cause of the Cottingley fairy photographs. Sir Arthur Conan Doyle, the celebrated writer and creator of the fictional detective Sherlock Holmes, was a believer in spiritualism. In June 1920 Doyle was commissioned by the *Strand Magazine* to write an article about fairies for its Christmas issue. He had heard of the fairy photographs and made contact with Gardner. At first Doyle was very guarded in his response to the photographs. He showed them to

an eminent psychical researcher, Sir Oliver Lodge, who judged them fakes. (Lodge remarked on the somewhat contemporary bobbed hairstyle and clothing of one of the fairies.)

A committed spiritualist
However, Doyle chose to accept the pictures at face value: they showed actual fairies. And such was his reputation in early 20th-century England that his support for the photographs gave them a credence that they would not otherwise have enjoyed. Doyle was a committed spiritualist; he also believed that if the photographs were accepted as proof of the existence of fairies, then other psychic phenomena would find a readier acceptance among the public.

Epoch-making event
Doyle published a first article in July 1920 in the *Strand Magazine*, using Gardner's two enhanced prints under the headline: FAIRIES PHOTOGRAPHED. AN EPOCH-MAKING EVENT DESCRIBED BY A. CONAN DOYLE. The issue was an immediate sell-out, and the article became the subject of national debate.

In August 1920 Gardner, at Doyle's request, went to Cottingley to talk to the girls. He brought with him a new camera and some 20 specially marked photographic plates, as he wished to persuade the girls to take more pictures. The girls obliged and produced three more photographs on August 17, 1920. These consisted of a leaping fairy watched by Frances, a fairy offering a bunch of flowers to Elsie, and two almost translucent fairies nestled on the ground among bramble bushes.

Debate down the decades
The Cottingley fairy debate continued to rage over the next 60 years. Those who believed in the photographs argued that it was unlikely that two unsophisticated girls could possibly possess the technical skill to create such photographic tricks. But Elsie had been apprenticed to a photographic studio for several months during the First World War. Here she might have learned how to retouch and even fake photographs. She was also a proficient artist, having attended drawing school in Bradford. Investigators in 1920

also seem to have disregarded the words of Elsie's mother that her daughter was "a most imaginative child, who had been in the habit of drawing fairies for years." So the years rolled by until Elsie and Frances were interviewed on a BBC television program called *Nationwide* in 1971. The cousins stuck to their original story: that they had seen fairies regularly in the Cottingley glen.

Illustrative evidence

Then in 1977 British writer Fred Gettings was researching early 20th-century illustrated children's books and found *Princess Mary's Gift Book* (1915). In this text he found an illustration of dancing fairies by Claude A. Shepperson, one that matched uncannily the dancing fairy figures in Elsie's first photograph. The only difference between the drawing and Elsie's photograph was that in the latter the fairies' wings were larger. This book was a bestseller in 1915, and Gettings established that Frances had owned a copy of this book.

Faked photographs

Paranormal skeptic James Randi revealed in his book *Flim Flam* (1980) how he had analyzed the Cottingley fairy photographs, using computer image-enchancement techniques developed by William S. Spaulding of Ground Saucer Watch, Phoenix, Arizona, an organization set up to test the authenticity of UFO photographs. This analysis revealed clear evidence of fakery: in the 1920 picture of a fairy offering a flower to Elsie, a wire could be seen. All the photographs were enhanced to establish

if the fairies were three-dimensional images, and every one failed the test.

The final piece of the Cottingley fairy puzzle was supplied by sociologist Joe Cooper, who was a friend of the two cousins. He had often asked them about the fairy photographs, and in 1982 the cousins — who were by then old ladies — revealed to Cooper the truth: that they had faked the photographs. They had drawn the figures and cut them out. They had "hid them in their bosom" and carried the cutouts to the beck. There they had carefully positioned them with hatpins and had taken the pictures. Afterward they had thrown the paper fairies into the fast-flowing stream, thus removing every trace of their deception.

> ### "In my mind these pictures were altogether beyond the possibility of fake."
> **Arthur Conan Doyle**

The final word

Elsie claimed that all five photographs were faked. Yet Frances, even to her dying day, insisted that the fifth photograph was real. In her own words: "It was a wet Saturday afternoon and we were just mooching about with our cameras and Elsie had nothing prepared....I saw these fairies building up in the grasses and just aimed the camera and took a photograph." Frances maintained that she had often seen fairies at the beck, and that the photographs were to prove this to the doubting adults.

The final word rests with Frances's daughter Christine, who spoke to Joe Cooper in 1986. "She [Frances] had often seen fairies. The last one popped up beside her...in the kitchen in Nottingham during the Second World War. She never changed her story about seeing fairies at Cottingley. They were real to her."

Arthur Conan Doyle

FAIRY BELIEVER

Why should the celebrated author Sir Arthur Conan Doyle be so willing to believe that the Cottingley fairy photographs were genuine? Arthur Conan Doyle was the eldest son of Charles Altamont Doyle. An architect and draftsman by profession, Charles Doyle was a highly strung and mentally unstable man who spent the latter part of his life in and out of mental asylums.

A fantasy world

Fairies were familiar inhabitants of Charles Doyle's fantasy world. In 1978 an English art dealer, Michael Bond, published a hitherto unknown collection of Charles Doyle's drawings, which revealed his troubled state of mind. In post-Victorian England an enormous social stigma was attached to those who were mentally unbalanced. Kevin I. Jones, in his biography *Conan Doyle and the Spirits* (1989), has concluded that Conan Doyle's predisposition to believe that fairies actually physically exist was a means of establishing that the "little people" his father saw were not purely the product of his hallucinations. For Conan Doyle, this would mean his father might not be considered insane.

Spiritual reunion

Toward the end of his life spiritualism became increasingly important to Conan Doyle. One reason to explain this was that his son Kingsley was killed in the trenches during the First World War. Some commentators have argued that spiritualism's increasing appeal at that time was a direct result of the tragic carnage of that war.

Spritualism offered a growing number of bereaved people some form of contact with those who had died. Conan Doyle, as a grieving parent, might well have hoped to find comfort in the notion that he might be reunited spiritually with his son.

SPONTANEOUS IMAGES

On eyes and eggs, on walls and windows, recognizable images have often appeared without human assistance. But any meaning they may have is ultimately perceived only in the eye of the beholder.

N 1951 THE BLUE EYES OF PETER JACOBS, a 19-month-old South African boy of Heidelberg, Transvaal, were reported to be marked in an extraordinary way. On the iris of his left eye could be seen the letters of the alphabet, and on the iris of his right the numbers 1 to 12. Johannesburg eye specialists who examined the boy declared that the phenomenon was the result of a billion-to-one chance in the formation of the eye pigment.

Remarkable eyes

Images on eyes are nothing new — though they are extremely rare. The Dutchman Henry Kens (born *c.* 1693), for example, was said to bear on the gray iris of his left eye the Latin phrase *DEUS MEUS* (*My God*) and on that of his right eye the Hebrew word *ELOHIM* (*God*). Kens's parents exhibited their son all over Europe.

However, do such eye markings genuinely form the words they are claimed to represent? In the case of Kens, some who observed his eyes did not think so. The Reverend Charles Ellis, an Englishman who saw Kens in Belgium in 1699, admitted in a letter he wrote to a friend that the filaments of darker pigment on the left iris "might be thought to form some imaginary letters....There is something like a D, and I, and V....But," he went on, "there is not a trace for the strongest fancy to work out any more, nor any letter of Hebrew in the right eye."

Growing characters

On the other hand, eight years later, the English writer James Paris du Plessis, who had also seen the boy, declared, in his *A Short History of Human Prodigies and Monstrous Births*: "Upon each eye...there was plain to be read in fair characters, like prints, Deus Meus. As he grew bigger, the characters grew larger and more legible." Du Plessis was obviously completely convinced by what he had seen — but it should be noted that what he claimed to have observed on the right eye differed from what other witnesses had perceived.

The celebrated English diarist John Evelyn summed up the conflict of opinion succinctly: "Physicians and philosophers examined him [Kens]...some considered it as artificial, others as almost supernatural."

Napoleonic inscriptions

Several examples of eye imagery occurred in France. One notable case, in the early 19th century, was that of a girl who allegedly had on her pale blue eyes images of the inscriptions that appear on a Napoleonic franc piece: on

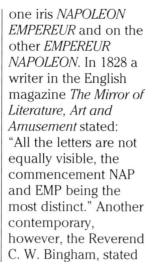

> ## Mrs. Gertrude Smith claimed that she could mentally coax her hens to lay pictorial eggs.

Comet egg
This contemporary engraving illustrates a bizarre phenomenon that was reported in Rome. In December 1680, when a giant comet appeared in the sky, a hen apparently laid an egg bearing an image of the comet against a background of stars.

"Spirit" on film
In 1973, after making a cine film of several TV shows, Stella Lansing of Massachusetts claimed to have found the film overlaid with mysterious images, among them this "spirit photograph" of an unknown man. Claimed images of spirits of the dead have been appearing on otherwise normal film since the early days of photography in the mid-19th century. Upon investigation, however, most have proved to be fraudulent.

one iris *NAPOLEON EMPEREUR* and on the other *EMPEREUR NAPOLEON*. In 1828 a writer in the English magazine *The Mirror of Literature, Art and Amusement* stated: "All the letters are not equally visible, the commencement NAP and EMP being the most distinct." Another contemporary, however, the Reverend C. W. Bingham, stated that same year in a letter: "The whole thing is a humbug...I had pictures and models of her eyes shown me, that I might know where to find the respective letters. Not one could I see!"

Pigment patterns
That some people should see some recognizable image in the darker pigment pattern in the iris may well have been determined simply by how imaginative, or gullible, they were: those who wanted to see letters or numerals saw them; those who could not sufficiently suspend their disbelief saw nothing meaningful. In other words, as with so many other apparently spontaneous images, seeing them or not probably depends entirely on the eye of the beholder.

Images on eggs
There are a number of cases of eggs mysteriously bearing images. In 1940 Mrs. Gertrude Smith of York, Pennsylvania, claimed that five years earlier she had discovered she could mentally coax her hens to lay pictorial eggs. "I would stand near the hen yard," she told the *York Gazette and Daily*, "and visualize sunflower petals along with my initials. In a few days my father came

into the house all excited and said: 'Here is the sunflower egg.' The pattern of a sunflower was incised into the shell on the flattened part." And she claimed, on another occasion, that her own initials, reversed, had appeared on another egg.

The appearance of comets in the sky is supposedly accompanied by the laying of "comet eggs." Halley's comet, which appears every 76 years, is particularly associated with the phenomenon. In 1910, a Halley's comet year, a county clerk named Fogg of Reno, Nevada, went outdoors to see the comet and found that his pet hen had laid "an egg with a long tail on it." In 1986 a hen owned by Linda Franklin of Studley, England, also produced a comet-shaped egg, which won the $10,000 first prize in a comet-egg competition that was organized by a leading egg producer.

Ghostly portraits
Apparently spontaneous images occur more commonly on buildings than on people. In 1923 there appeared on a wall of Christ Church cathedral, Oxford, some damp stains resembling the profile of a man's face. When a photograph of this was published in the London *Daily Express*, some Oxford readers apparently recognized the profile as that of Dean Liddell (father of Alice Liddell, model for the heroine of Lewis Carroll's *Alice*

books). Dean Liddell was an Oxford cleric who had died 25 years previously. Earlier, in 1897, the posthumous image of another cleric, Dean Vaughan, of Llandaff cathedral in Wales, had appeared mysteriously on a wall — this time in the form of a patch of mold on the cathedral wall. Charles Fort, noted American investigator of the bizarre and unusual, speculated that perhaps the portrait may have resulted from "intense visualizations of him [Vaughan], by a member of his congregation."

Death's heads

Not all the images said to have appeared spontaneously on buildings depict people. In 1872, in the town of Boulley, France, death's heads, eagles, rainbows, and other astonishing emblems allegedly materialized suddenly on the windows of houses. One window bore an image of Zouaves (a type of French infantrymen) waving banners. This alarmed the German authorities in nearby Metz, a French city that had been under German occupation since the Franco-Prussian War of 1870–71. They sent a detachment of troops to smash the window — but it was said that the Zouaves could still be seen amid the shattered glass.

Irremovable images

About 100 miles away, in Baden-Baden, just over the German border, the bizarre phenomenon was repeated: the bemused inhabitants awoke to find crosses printed on their windows. Yet no amount of washing, even with acids, could remove them. How these images were created has never been established, but their interpretation as vaguely threatening is perhaps attributable to the paranoia of the inhabitants of a recent war zone.

Mystery fresco
In a house in the Spanish village of Belmez de la Moraleda, in 1971, this and other inexplicable portraits of unknown people were said to keep reappearing on the kitchen floor, even when it had been cemented over.

THE TURIN SHROUD

In a silver casket in the Royal Chapel of Turin Cathedral, Italy, is sealed a fragment of cloth that many believe to be Christ's burial shroud; a shroud declared by Pope Paul VI in 1973 to be "the most sacred relic in all of Christianity." This strip of linen, 14 feet long by 3 1/2 feet wide, bears two faint, straw-colored images of a naked, bearded man, one showing the front of the body, the other the rear. On the head, trunk, wrists, and feet are slightly darker, rust-colored stains, said to be of blood and consistent with crucifixion wounds. Those in a ring around the forehead are suggestive of smaller puncture wounds. The back bears marks that might indicate a flogging.

Startling negative

The Turin Shroud first came to light in 1357, when it was exhibited in a church at Lirey in France. It then passed into the hands of the dukes of Savoy before being brought to Turin in 1578. Modern scientific interest in the shroud began in 1898 when Secondo Pia, an Italian, took the first photograph of it. This revealed for the very first time the shroud's most remarkable characteristic: the image on it, which shows strange, murky facial and bodily features, is identical in tone to a photographic negative. Pia realized this when his own negative showed a tone-true, lifelike portrait, which normally only a positive would — and one, too, which depicted a majestic Christ. Pia was so startled that he dropped the negative plate.

This revelation prompted an arresting question: if the shroud was a forgery, as had

often been claimed, how could a medieval artist have anticipated the discovery of photography centuries later to achieve this striking effect?

One highly speculative theory designed to explain the "negative" effect, put forward by those who believed the shroud to be authentic, was that the image was formed by a burst of radiation from Christ as he lay in His tomb. Thus the image became, in the words of English writer Ian Wilson, "a literal 'snapshot' of the Resurrection." Skeptics have offered more straightforward explanations, for example, that the image was produced by scorch marks or by a rubbing of a bas-relief sculpture.

Scientific testing

During this century the shroud has probably undergone more scientific testing than any other relic in history. The linen, and pollen found in it, is considered definitely to have originated in the Middle East. Physicians have testified to the anatomical accuracy of the body image on the shroud and to the authentic size and disposition of the "bloodstains."

Then, in 1988, three laboratories that had each separately carried out radiocarbon dating on fragments of the shroud announced that the fabric fibers might be dated to between 1262 and 1384. Yet there is a considerable margin of error in radiocarbon dating; in previous datings of other material, two of the three laboratories had been wrong by several centuries. And so, unless more conclusive evidence is produced, many will continue to believe that the shroud is indeed a relic of Christ's crucifixion.

HATS, CATS, AND LADDERS

Every time you touch wood for luck, you are participating in an ancient belief that there is a spirit in the wood that can protect you. From earliest times, humankind has had faith in the power of superstitions to summon good luck and keep bad fortune at bay.

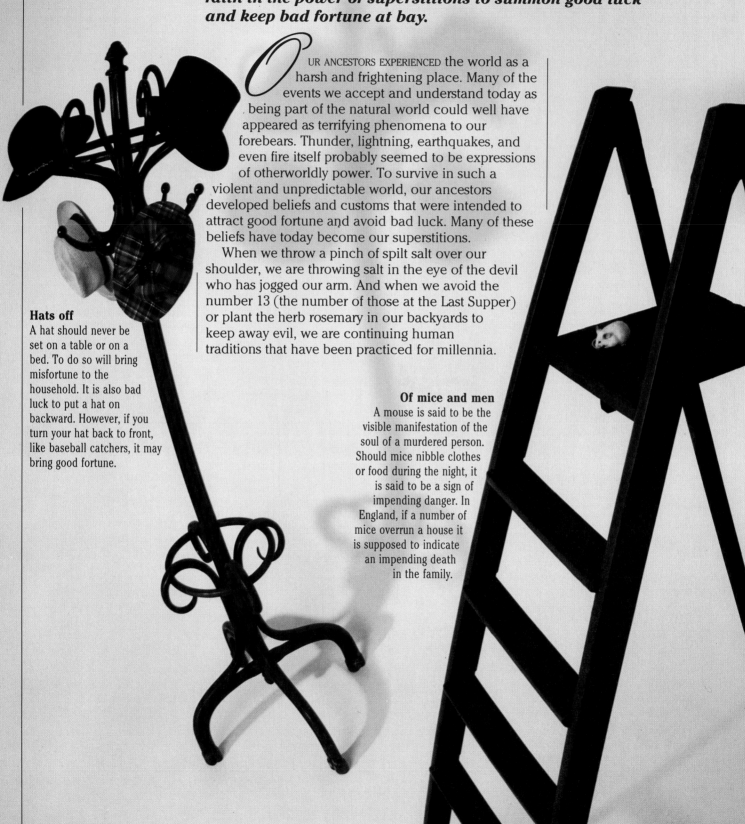

OUR ANCESTORS EXPERIENCED the world as a harsh and frightening place. Many of the events we accept and understand today as being part of the natural world could well have appeared as terrifying phenomena to our forebears. Thunder, lightning, earthquakes, and even fire itself probably seemed to be expressions of otherworldly power. To survive in such a violent and unpredictable world, our ancestors developed beliefs and customs that were intended to attract good fortune and avoid bad luck. Many of these beliefs have today become our superstitions.

When we throw a pinch of spilt salt over our shoulder, we are throwing salt in the eye of the devil who has jogged our arm. And when we avoid the number 13 (the number of those at the Last Supper) or plant the herb rosemary in our backyards to keep away evil, we are continuing human traditions that have been practiced for millennia.

Hats off
A hat should never be set on a table or on a bed. To do so will bring misfortune to the household. It is also bad luck to put a hat on backward. However, if you turn your hat back to front, like baseball catchers, it may bring good fortune.

Of mice and men
A mouse is said to be the visible manifestation of the soul of a murdered person. Should mice nibble clothes or food during the night, it is said to be a sign of impending danger. In England, if a number of mice overrun a house it is supposed to indicate an impending death in the family.

Rainmaker
It is bad luck to open an umbrella indoors. Originally designed as a protection against the sun, opening an umbrella anywhere where there is no sun will bring misfortune. If an umbrella is opened unnecessarily during fine weather, rain will soon follow. If an unmarried woman drops an umbrella and picks it up herself, she will remain a spinster all her life.

A mirror cracked
A broken mirror is commonly meant to bring seven years of bad luck, or some more specific misfortune, such as the loss of a close friend or a death in the family. In Scandinavia, it is said, a woman should never look at herself in a mirror by candlelight, for she will lose her beauty. According to Caribbean folklore, if the finger of a dead person is rubbed on a mirror, then the dead person's soul will be released from the body.

Apple of your eye
In folklore the apple has represented immortality, eternal youth, and happiness. If the sun shines through the boughs of an apple tree on Christmas Day, then fruit will be abundant the next year. If a young girl wishes to find the name of her future husband, she must twist the stalk of an apple, repeating the alphabet, with one letter for each turn. The letter at which the stalk breaks will be the first letter of her future husband's name.

Seat of fortune
Tipping over a chair is considered to be bad luck. If you tip over a chair after a meal, it means you are a liar. If when you leave someone's house, you push your chair against a wall, it means you will never visit that house again. In Ohio if three chairs are accidentally placed in a row, it indicates, according to the lore of superstition, that there will be a death in the house.

Black cat
In the U.S. it is unlucky if a black cat crosses your path. In Britain, however, this is lucky. If, when someone is ill, a cat leaves a house, and cannot be coaxed back, then that person will die. But if a bride leaving for her wedding hears the family cat sneeze, then she will have a happy marriage.

Divine triangle
It is bad luck to walk through the angle formed by a ladder against a wall, or with the ground. The triangle this forms is interpreted as representing the Holy Trinity, the three-in-one godhead of God the Father, Son, and Holy Ghost. To walk through this triangle is to show disrespect and it also leaves one open to attack by the devil. To counter this bad luck, the story goes, cross your fingers until you see a dog.

CURSES AND JINXES

Since ancient times, priests, sorcerers, and witches have called down curses upon those they wish to harm. Sometimes, it appears, the simple suggestion that a curse has been laid is enough to produce some devastating effect.

ON NOVEMBER 4, 1922, a British archeologist, Howard Carter, discovered the entrance to a sealed tomb in the Valley of the Kings near the city of Luxor in Egypt. Carter telegraphed his discovery to his patron, the earl of Carnarvon, who had funded his excavations in the Valley of the Kings for the last 14 years. Lord Carnarvon immediately departed for Egypt with his daughter, Lady Evelyn. In the dead of night, on November 26, Carter with his assistant Arthur Callender, Lord Carnarvon, and Lady Evelyn burrowed into the undisturbed tomb to gain entry to the burial chamber of the boy king, Tutankhamen, and discover its dazzling treasure.

Ill-omened bird

On his last visit to England, Carter had bought a yellow canary to celebrate Lord Carnarvon's agreement to continue funding the excavations. The Egyptian diggers said it would bring Carter luck. However, a matter of days after the tomb was opened, a cobra found its way into the canary's cage and ate the bird. The Egyptian workers saw this as a bad omen: the pharaohs were protected by the cobra goddess Wadjet, whose image can be seen rising from the burial headdress of Tutankhamen. The canary's sad end, they believed, signified the goddess's displeasure, and a death would soon follow.

Death among the pharaohs

Hieroglyphics that were found on a tablet above the entrance to the tomb were interpreted as bearing the curse: "Death will slay with his wings whoever disturbs the pharaohs' sleep." By the end of February 1923, only weeks after the official opening of the tomb, Carnarvon fell ill in Cairo. He died on April 4, 1923, apparently from blood poisoning as a result of a mosquito bite. During his final illness, a delirious Carnarvon had muttered many times: "A bird is scratching my face." At the precise moment of his passing, it was later discovered, Carnarvon's fox terrier, Susan, who was at Highclere Castle, his country estate in England, howled and died.

From these events grew the story of the curse of Tutankhamen. And further significance was attached to the number of archeologists and visitors who were taken ill or even died after entering the tomb. In 1923 an American multimillionaire, George Jay Gould, died suddenly of a fever after visiting the tomb. A Canadian friend of Carter's, Prof. M. Laffleur, developed a fever and died

> # "Death will slay with his wings whoever disturbs the pharaohs' sleep."

the day after his first visit to the tomb in 1924. Two of Carter's assistants, A. C. Mace and Richard Bethell, died in 1924 and 1928, respectively. By 1929, 22 people who had been associated with the excavation had died prematurely.

The curse explained

In 1980 a British television company was researching a documentary entitled *The Curse of King Tutankhamen.* Researchers tracked down an ex-military policeman, Richard Adamson, who had been employed to guard the tomb. He revealed that Carter had himself started the rumor of a curse to keep sightseers away and to deter thieves from entering the pharaoh's tomb.

The prime candidate to suffer the full force of the curse should surely have been Howard Carter. He never believed in the curse. Carter died of natural causes in 1939, aged 66.

Family curses

The effect of curses has been recorded since earliest times. In ancient Greek legend, when Atreus, the king of Mycenae, killed the

Opening the tomb
Howard Carter and his assistant Arthur Callender at the moment they uncovered Tutankhamen's stone sarcophagus.

THE PHARAOH'S CURSE
Theories abound to explain the curse of Tutankhamen. One theory was that the objects in the tombs were coated with some poison, an art in which the ancient Egyptians were notoriously skilled.

Another explanation is that bacteria that may have been flourishing for thousands of years within the tomb infected visitors. In 1949 atomic scientist Professor Louis Bulgarini argued that the floors might have been covered with uranium-bearing or some other type of radioactive rock.

Death by radioactivity
In 1991 Dr. Sayeed Mohammed Thebat, an Egyptian scientist from Cairo University, put forward a new theory. He too believed that the deaths of people entering unopened tombs were caused by radioactivity. This was released, he argued, by some as yet unidentified substance that was part of the mummification process. Dr. Thebat made his discovery when walking through a chamber of mummies in the Cairo Museum. He said he noticed that the Geiger counter he was carrying registered high radiation levels.

son of the god Hermes, the latter put a curse on Atreus and all his descendants. In the catalog of tragedy that followed, Atreus killed his own son; Atreus was then killed by another son; his grandson Agamemnon was murdered in his bath by his wife, Clytemnestra. She, in turn, was murdered by her son and daughter.

The Craven curse
In more recent times the continuing effect of some curses can still reportedly be seen. On August 30, 1990, Simon Craven, age 28, the eighth earl of Craven, was killed in a car crash in the English seaside town of Eastbourne. His death was the latest in a series of events that seemed to fulfill a 350-year-old prophecy that all the earls of Craven are doomed to die young. This curse, so the story goes, was laid upon the Craven family by a serving girl who was made pregnant, and abandoned by the second earl. In 1983 Lord Craven's elder brother Thomas, the seventh earl, shot himself in a fit of depression at the age of 26. Their father had died from leukemia at the age of 47. In 1932 Lord Craven's grandfather had died at the age of 35 after excesses at a wild party on board a yacht.

A jilted woman
In the 13th century Prince Rainier I, of the principality of Monaco on the French Riviera, was renowned for his amorous adventures. However, his luck ran out when he jilted a Flemish woman. According to the legend, in her distress the woman cursed the Grimaldi family, saying: "Never will a Grimaldi find true happiness in marriage." The marital history of the Grimaldi family in the 20th century appears to reflect this curse: Charlotte and Pierre, the parents of the current prince, Rainier II, divorced; Prince Rainier II himself was left a widower prematurely when his wife Princess Grace, the former movie star

Grace Kelly, died in a car crash in 1982; their daughter Caroline's first marriage failed and in 1990 her second husband, Stefano Casiraghi, was killed in a tragic powerboat accident.

Curses are still cast by priests of some of the major religions. In January 1991 the ultra-orthodox Jewish Eda Haredit group in Israel performed the "Rod of Light" ceremony in which a death curse was pronounced on Saddam Hussein. This rare ceremony was held in a room lit by black candles where at least 10 righteous men gathered to recite cabbalistic incantations, and burn a paper on which they had inscribed the names of Saddam Hussein and his mother. The *shofar* (ram's horn) was blown to dispel the *shedim* (evil spirits).

Persistent jinxes
While curses are deliberate invocations of misfortune against others, jinxes are patterns of bad luck that appear to lie well outside normal experience. For example, an extraordinary series of misfortunes has befallen members of the Guinness family. This Irish brewing family, whose fortune comes from the famous beer that bears their name, would on first inspection appear to enjoy enormous privilege: wealth, good looks, power, and influence. However, the family seems to have been jinxed by a series of fatal accidents, inexplicable suicides, and accidental drug overdoses.

In 1978 the family suffered four deaths in as many months. In May, Lady Henrietta Guinness, age 35, fell to her death from an aqueduct in Spoleto, Italy. In June, Natalya Citkowitz, a Guinness heiress, died in her bath from a drug overdose at 17. In July, Maj. Dennys Guinness, age 47, died from a suspected drug overdose. Finally, in August, Peter Guinness, age 4, whose father was an adviser to the then British prime minister James Callaghan, died in a car crash. And in 1986 Olivia Channon, age 21, a member of the family, died from an accidental drug overdose.

Jinxes can be very persistent. Doreen Squires had reason to be apprehensive

> ## "Never will a Grimaldi find true happiness in marriage."

HEADS OR TALES?

In February 1972 two young boys, Colin and Leslie Robson, unearthed a pair of small, heavy stone heads in the backyard of their house at Hexham, in Northumberland, England. One of the heads was skull-like and masculine, the other witch-like. (Colin Robson claimed he had constructed a vaguely similar head shortly before.) The boys took their finds into the house — and then, allegedly, strange things began to happen. The heads supposedly turned around of their own accord, and a luminous flower grew in the backyard where the heads had been found.

Large black creature

Eventually, a Celtic scholar, Dr. Anne Ross, of Southampton University, took possession of the heads. Dr. Ross identified them as Celtic ritual sculptures carved of local Northumbrian stone during the 2nd century A.D. One night she woke up feeling cold and frightened and in the doorway saw a tall, black, furry, werewolf-like being. She followed it downstairs and saw it disappear near the kitchen. A few days later Dr. Ross's teenage daughter had a similar experience. Sometimes, Dr. Ross recounted, her family also heard soft padding footsteps.

Dr. Ross removed the heads from the house, but she was still aware of a disturbing presence, which she associated with the other Celtic heads

Mystery heads
On the left is the head that Desmond Craigie claimed he made, on the right the one that Colin Robson allegedly made before finding the so-called Celtic heads.

in her collection. Only when she had disposed of her entire collection of Celtic heads did she feel that things had returned to normal.

It seemed that the Celtic heads might possibly have carried some ancient curse. The Celts are known to have performed human sacrifices and worshiped half-animal gods in front of carved stone heads of the kind found at Hexham.

Lost playthings

However, later in 1972, Desmond Craigie, a truck driver who had previously occupied the Robsons' house, announced that he had made the heads from cement 18 years earlier. His daughter had lost them while playing in the garden. Dr. Ross replied that they were indisputably Celtic in style, could not have been made without considerable study of similar heads, and had been shown by analysis not to contain calcium silicate, an essential constituent of cement. Even so, after Desmond Craigie's claim, Dr. Ross was inclined to keep an open mind about the true date of origin of the Hexham heads.

The two heads have since disappeared, and the mystery surrounding them remains. Were they genuinely Celtic, and did they have some malign influence? Or were they crudely made modern objects whose alleged effects were like those of a poltergeist?

---

when a piece of steel flew into her 25-year-old son Martin's eye while he was working at a sawmill in Totnes, Devon. Her father, grandfather, and great-grandfather had all lost their right eyes in accidents. Doreen's great-grandfather, Jim Chapple, who was a stonecutter, lost his eye when it was injured by a stone chipping. His son, Jim, a blacksmith, lost his eye when a piece of steel flew up as he was forging a horseshoe. Doreen's father, Adrian, lost his eye in a quarry blast. All three had been born on the same day, September 29. Martin, however, recovered his sight. Why? Perhaps because Martin Squires was not born on the same day.

An unlucky engine

Some inanimate objects seem to carry their own jinx. Locomotive D326, built by English Electric in 1960, was one of the first mainline diesels to replace express steam engines. In 1962 the locomotive was involved in a collision with another train north of Crewe, England, in which 18 people were killed and 33 injured. The locomotive was repaired and put back

to work on the west coast mainline. Within months, the engine's picture was on the front pages again: D326 was waylaid by thieves in August 1963. The driver was severely injured, and a record haul of $5 million was taken in what became known as the Great Train Robbery.

Off the rails
Locomotive D326 has had a checkered history. In 1962, the day after Christmas, the train crashed causing numerous fatalities. Eight months later the same locomotive was involved in a robbery in which the thieves got away with a haul of over $5 million.

CHAPTER FIVE

LIVING LEGENDS

Human magnets, living conductors of electricity, mythical giants and dwarfs, and lost tribes — are the stories that describe such strange people just make-believe, or might they be true reports of the extremes of which humankind is capable?

The Tenkaevs of Saratov, Russia, are a very unusual family: Leonid Tenkaev, a factory worker, his wife Galina, their daughter Tanya, and Tanya's son Kolya reportedly all have the uncanny ability to make metal objects stick to them. According to Dr. Valeri Lepilov, professor of physics at Saratov State University, the four have only to "concentrate and think about generating heat inside their bodies" in order to trigger their alleged magnetic mechanism, which is sometimes very powerful. Leonid, for example, who was born in 1928, reportedly can cause up to 52 pounds of ferrous metal (containing iron) to stick to him at any one time. Removing it afterward, according to Dr. Lepilov, is often very difficult — "like dragging a metal object off a real magnet." When the family was flown to Japan to appear on a television program, their talents were witnessed by Dr. Atsusi Kono, chief physician at the Djo Si Idai Hospital in Tokyo, who was prompted to comment: "There is absolutely no doubt that the objects stick as if their bodies were magnetic."

Nuclear link

It appears that the Tenkaevs first noticed the phenomenon in 1987, the year after the nuclear accident at Chernobyl, which is about 700 miles to the west of their hometown of Saratov. The sudden appearance of the phenomenon suggests that it might be linked to the disaster in some as yet unexplained way.

However, the Tenkaevs are not the only eastern Europeans whose bizarre abilities have suddenly appeared in the news since the Chernobyl nuclear disaster. For example, Russia's *Soviet Weekly* carried an intriguing story in June 1990 about a militia patrolman named Nikolai Suvorov who was also able to make metal objects stick to him. Then in 1991 Bulgaria's Sofia Press Agency reported that no fewer than 300 "magnetic" people turned up at a contest to see who could keep assorted metal objects on their bodies for the longest time.

The growing number of reports of this kind is just as likely to result from the increase in press freedom during recent years as from some unknown and disturbing effect related to nuclear fallout. However, if these abilities are valid, it seems likely that they are associated with an electrical force of some type, since magnetism and electricity are closely connected. Yet the precise means by which they might be triggered still remains a mystery, as is the way they appear to be turned on and off at will by the people involved. Even more curious is the fact that the effect is not always a magnetic one in the strictly scientific sense — unlike ordinary iron magnets, human magnets are supposedly not limited to attracting ferrous metals.

Militiaman Suvorov, for instance, attracts plastic and glass as well as metal, while a teenager named Inga Gaiduchenka from the city of Grodno, Belorussiya, claims that she can attract metal, plastic, wood, and paper — but not glass. Something more like gravity than magnetism might appear to be at work here. But then why, in the latter case, should glass escape its attention?

Little Georgia Magnet

A related ability was reportedly demonstrated by a young American girl at the end of the 19th century. Annie May Abbott, dubbed Little Georgia Magnet, weighed 98 pounds, yet when she sat in a chair, the downward force she apparently exerted was so great that even Sumo wrestlers could not shift her. She was, it seemed, able to make objects immovable simply by resting her fingers on them.

For some people, the powerful effect of electrical energy in their bodies is seemingly much more direct. At the

> **In 1991 Bulgaria's Sofia Press Agency reported that no fewer than 300 "magnetic" people turned up at a contest to see who could keep metal objects on their bodies for the longest time.**

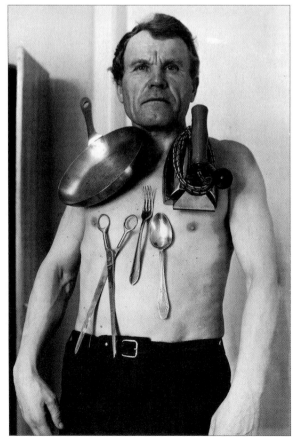

Metal magic
Russian militia patrolman Nikolai Suvorov was 55 years old when he discovered his strange magnetic abilities.

beginning of 1988, Xue Dibo, a boilermaker from Urumqi, in China's Xinjiang province, began to feel what he described as "very strange sensations." Suddenly, when he touched people, he discovered that he was apparently giving them an electric shock strong enough to knock them over.

Static woman

Another case was researched by Dr. Michael Shallis of Oxford University in 1985. Shallis discovered that Mrs. Jacqueline Priestman of Sale, Manchester, England, had more than 10 times the normal amount of static electricity in her body, a fact that seemed to explain the destructive effect she had on her household equipment. This effect first manifested itself, for no apparent reason, when she was 22 years old. Since then, she has had to purchase a staggering 30 vacuum cleaners, five electric irons, and two washing machines.

Changing channels

When she approaches her television, it reportedly changes channels; when she plugs in her kettle, sparks fly out of the socket; when she turns on her electric stove, it cuts out. According to Dr. Shallis, "It's not certain what

Healing strength
As well as attracting metal objects, Belorussiyan teenager Inga Gaiduchenka can alleviate joint pain with the unusual warmth of her hands.

Power source
In this dramatic photograph, Brian Williams from Cardiff, Wales, demonstrates his claim to be able to illuminate a light bulb just by holding it between his fingers.

causes this severe build-up of static. Such people can, it seems, transmit miniature bolts of lightning that break down the insulation of some electrical appliances."

Electric build-up

But Shallis's theory does not explain why the build-up began when Mrs. Priestman was 22, or why its effects are limited to her own home.

In itself, there is nothing strange about electricity in the human body: without minute electrochemical charges making connections in the brain, we could not think or move. But human beings do not normally have the capacity to store and discharge electricity. Electric eels, however, can deliver a jolt of up to 500 volts. The existence of these eels suggests that electric people may not be breaking the laws of nature, even if the cause of this ability is not yet fully understood.

UNSHOCKABLE

The ability of certain individuals to generate electricity is certainly unusual, and often disturbing for its victims. Yet perhaps more astounding is some people's apparent capacity to cope with electrical currents in powerful and dangerous doses.

Shock sensation

In May 1984, reports reached the western world about an 80-year-old Bulgarian electrician named Georgi Ivanov. According to Dr. Georgi Thomasov, writing in the *Bulgarian Medical Journal*, Ivanov had a natural talent that was very useful in his profession — he was immune to electric shocks. When he was on a job, Ivanov never turned the power off, since he considered it too much trouble. When subjected to the full force of the electricity at the main service panel (13 amps at between 220 and 240 volts — enough to be lethal), all he felt was a tingling sensation. When tested at a Sofia hospital, Ivanov was reported to have absorbed 380 volts at standard domestic amperage before he felt any discomfort.

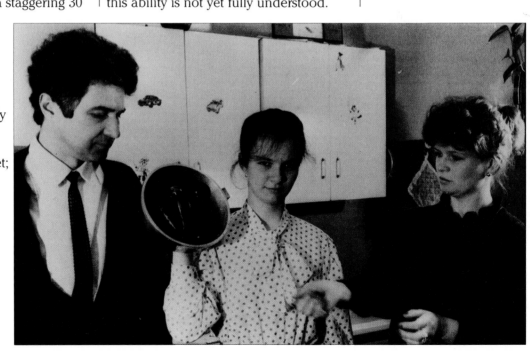

GREAT AND SMALL

We have read about giants and dwarfs in fairy tales. And there is evidence that such people once lived on the earth.

IN A CAVE in Caspar, Wyoming, in October 1932, two prospectors found the mummified figure of a middle-aged man who was reportedly only 14 inches high. The Anthropology Department at Harvard University reportedly declared that the remains were of human origin. In the 1880's in Sayre, Pennsylvania, excavations of a prehistoric mound revealed skeletons of 68 men. The average height of the skeletons was apparently seven feet, with many of them taller. These finds suggest that there used to be a greater variety in human size than now exists.

Legendary creatures

Great and small people have become incorporated into the legends of many cultures. In the Bible, Genesis 6:1–4, the "sons of God," or fallen angels, produced giants by interbreeding with the women, "the daughters of man," who lived beyond the Garden of Eden. In Greek legend, giants issued from the union of Gaia (earth) with Uranus (sky). In Ireland there is a folklore tradition that the Little People, or fairies, were a race of small people who hid under the ground from the invading Celts, in about 500 B.C. Our genetic inheritance regularly produces genes that create very tall or tiny people. Is it possible that they are throwbacks to a time when giants and dwarfs walked the earth?

Giant at the gates

Gog, with his brother Magog, were the last British giants, according to Geoffrey of Monmouth in his *History of the Kings of Britain* (*c.* 1135). Brutus, the legendary founder of Britain, reportedly forced them to act as porters at the gates of his palace in London. Statues of Gog, seen here, and Magog stand in the City of London's Guildhall building.

Tom Thumb Charles S. Stratton, from Bridgeport, Connecticut, never grew taller than 35 inches. He became famous worldwide traveling with Phileas T. Barnum's circus, where he was known as General Tom Thumb. He is dressed here as Napoleon Bonaparte.

118

The Long Man of Wilmington
The largest representation in the world of the human form — 226 feet from head to toe — is cut out of the chalk hillside near Windower, Sussex, in the south of England. Some people interpret it as showing a Saxon warrior god. Others believe the figure marks the spot where one legendary giant killed another in a quarrel.

Egyptian dwarf
The ancient Egyptians believed that dwarfs were survivors of races of tiny men. This statue from the Fifth Dynasty (*c.* 2500–2350 B.C.) shows Seneb, chief of all the dwarfs, whose duty it was to look after the pharaoh's clothes, with his normal-sized wife. The couple's two children are depicted at the base of the statue.

UNHAPPY GIANTS
A tumor of the pituitary gland, located in the center of the brain, is the usual cause of gigantism, or excessive growth. The pituitary gland controls the levels of growth hormones in the body. Extreme growth is the result of an excess of the growth hormones, and such a condition may be hereditary.

Life of pain
Such extreme height can cause the individuals concerned great distress. The four O'Brien brothers were giants who lived in 19th-century Ireland. They apparently lived in horror at the thought of scientists dissecting their bodies after their deaths.

The tallest known giant was a 19th-century American, Robert Wadlow of Illinois. By the age of 10, he had reached 6 feet 5 inches. He was 8 feet 11 inches when he died at the age of 22. The speed at which he grew meant that he lived in constant pain.

The king of the fairies
In many folktales the enchanted, miniature world of the fairies is described as having a similar social hierarchy to that of the full-size, human world — complete with a king and queen.

Pygmy survivors
The Mbuti pygmy tribe has lived for centuries in the Ituri Forest in what is now Zaire, West Africa. The average stature of the men is 4 feet 6 inches. Anthropologists have estimated that some 40,000 pygmies survive in this area of Africa.

THE CUNNING LEPRECHAUN

One night the Irish poet William Butler Yeats and two friends were walking along a beach in the west of Ireland when "the unreal... [began] to take upon itself a masterful reality," and he found himself surrounded by a host of fairy folk. Yeats questioned the "tall queen" of this band of trooping fairies until "at last she appeared to lose patience."

The little people

Fairies in Ireland are always referred to as the Little People. This is a fairly direct translation of the Gaelic word *luchorpan*, meaning "small body" from which the modern word "leprechaun" probably comes. The little people, as Irish historian Brendan Lehane points out, are believed to be the magical remnant of the legendary *Tuatha de Danaan*, the people of the goddess Danu. They ruled Ireland before the invading Celts, in their turn, conquered them. These legendary people were then relegated to a low status in the supernatural world, Lehane writes, but retained certain of their former characteristics; chiefly an obsessive cunning.

Inventive liars

In Irish folklore, leprechauns are regarded as solitary creatures who guard stores of gold that are hidden in crocks. To find the leprechaun's gold you must first catch one. This is easier said than done. To even catch hold of a leprechaun, the little man has to be crept up on unawares, usually as he is working at his traditional craft of shoe-making. Then you hold on to him tenaciously. You must resist the blandishments of the silvery-tongued leprechaun who will try to keep the location of his gold safe. If his beguiling words momentarily cause you to let slip your grip of him, he will flee. This is why leprechauns have a reputation for being inventive liars.

Lucky leprechaun
This leprechaun, in the form of a silver lucky charm, holds a four-leaved clover for added good fortune.

In the land of the giants

In *Gulliver's Travels* (1726), the Irish novelist Jonathan Swift tells the fabulous tale of a sailor, Lemuel Gulliver, who travels to the land of Lilliput where the inhabitants are only six inches tall. When another voyage takes him to the land of Brobdingnag, Gulliver finds the inhabitants are giants. This tale was originally written as a biting satire on 18th-century England.

Giant builder

An astonishing promontory that stretches half a mile out to sea off the coast of County Antrim, in the north of Ireland, is known as the Giant's Causeway. According to legend, it was built by the Irish giant Finn MacCool. However, geologists believe volcanic activity during the Tertiary period of geological history may have thrown up lava to form the distinctive hexagonal and pentagonal columns of closely packed, gray-colored basalt.

Dwarf frieze
These three ugly, bejeweled dwarfs decorate the walls of a temple in Polonnaruwa, Sri Lanka. According to Hindu tradition, dwarfs were believed to be protective spirits, bringing good luck and wealth.

God of wealth
In Hindu mythology, Kuvera is the god of wealth. He has always been depicted as a grotesque dwarf with a paunch, and wearing jewels. Kuvera began life as a thief, but through many incarnations he grew virtuous and was included in the Hindu pantheon.

David and Goliath
In the Bible (1 Samuel 17), David slays the Philistine giant Goliath in single combat with a stone launched from his sling. He then takes the Philistine's sword and cuts his head off, as shown in this 15th-century illuminated manuscript. This story is an archetype of a theme found later in fairy tales: The hero overcomes a wicked giant, and so restores order and harmony to the world.

LOST TRIBES

Like giants and dwarfs, lost peoples may not have completely disappeared. They are more likely to have interbred and been assimilated into the communities in which they found themselves.

IN THE SUMMER OF 1984 A RAGGED BAND of emaciated men, women, and children trekked for 20 days across the blistering desert of drought-stricken Ethiopia in northwest Africa. Disease and starvation claimed many of their number; and only the strongest survived the journey. These refugees were the Falashas, or Ethiopian Jews; people who had been cut off from the rest of Judaism for over 2,000 years. The Falashas — the word means foreigner, or wandering one, in the Ethiopian language — claimed that they were descended from the tribe of Dan, one of the legendary Lost Tribes of Israel.

Promised homeland
In this desperate time of famine the Falashas were fleeing to the Sudan in the hope of emigrating to Israel. In what *The Times* of London called "one of the most dramatic migrations in the history of the Jewish people," the Falashas were airlifted from the Sudan and resettled in Israel. This migration had first begun in 1975, two years after the government of Israel had finally accepted that the Falashas were Jews and therefore welcome to return to Israel. (Israel's Law of Return provides that every Jew can emigrate to Israel.) By the spring of 1992, virtually the entire population of 40,000 Falashas had been resettled in their long lost homeland of Israel.

The Lost Tribes of Israel
According to the Bible, in the 10th century B.C., a revolt in the House of David split the kingdom of Palestine into two warring factions. Ten tribes lived in the north, Israel: while two others remained loyal to King Rehoboam (son of Solomon) and lived in the south, Judah. The two kingdoms fought each other over the next 200 years. In 721 B.C. the Ten Tribes of Israel were conquered by the Assyrians and forced to abandon their homeland for Assyria, where they seemingly disappeared. As the Old Testament notes in 2 Kings 17:23: "So Israel was exiled from their own land to Assyria until this day."

From there the trail grows cold. Aside from a few somewhat vague references to the Ten Tribes of Israel in ancient accounts, no one knows what became of them. Persecution of the Jews was common during the period of the Crusades in the 12th and 13th centuries.

At this time there were rumors of a vast Jewish empire in Persia, present-day Iran, that had been founded by descendants of the Ten Tribes of Israel. Although European Jews were encouraged to flee to this fabled land, there are no historical records of its existence.

Indian tribes
Nevertheless, in his 17th-century book *The Hope of Israel*, Rabbi Manasseh ben Israel, from Amsterdam, Holland, claimed that a Spanish-Jewish traveler had met South American Indians who spoke Hebrew and practiced Jewish rituals. Other explorers returned from their travels to the New World with similar claims. Mayan Indians and other American Indians in Peru and the Yucatán were believed by these travelers to be descended from the Lost Tribes. During the early exploration and settling of North America, interest increased in the fate of the lost tribes, and many believed that American Indians were their direct descendants. Some claimed that many of the American Indians' customs and folklore, such as circumcision, belief in flood myths, sacrifices to gods, and the veneration of a tribal ark, were similar to Jewish practice. In the late 1700's Scottish settler James Adair, who traded in Indian territories, wrote that he heard Choctaw and Chickasaw Indians use words very similar to certain words in Hebrew.

The hypothesis lives on with the Mormon Church. The Book of Mormon is considered by some to be a religious history of one of the Lost Tribes that settled in North America around 2000 B.C.

For many people, however, the Lost Tribes theory is nothing more than a fanciful legend. Most scholars believe that the Ten Tribes were not "lost" but merely assimilated into the peoples of various regions of the Middle East when they were banished from Israel. As a consequence, as they lost their identity, suggest the experts, the Ten Tribes disappeared completely from the pages of history.

Travelers from Wales
History is dotted with other instances of "lost" peoples. One of the most famous of these legends grew around the 12th-century Welsh prince, Madoc ab Owain

> **Rabbi Manasseh ben Israel, from Amsterdam, Holland, claimed that a Spanish-Jewish traveler had met South American Indians who spoke Hebrew and practiced Jewish rituals.**

Wandering tribe
In December 1991 these Falashas walked for over two weeks from Quara in northern Ethiopia to an assembly point at Atzdamaryan. From here, a Jewish agency took the Falashas on the final stage of their journey to Israel.

Gwynedd, who was reputed to have fled his homeland with 120 followers for North America. According to this story, first published in 1584, Madoc and his settlers reached Alabama in 1170. This would make Madoc and his followers the first Europeans to settle in North America, 300 years before Christopher Columbus landed in the West Indies.

Blue-eyed braves

Although there is little to back up the story, many subsequent settlers and explorers — including English seafarer and explorer Sir Walter Raleigh — claimed that American Indians spoke words derived from Welsh. By the 1800's there had been scores of such reports, from throughout the eastern half of North America. Even the early pioneer Daniel Boone claimed that a tribe of "blue-eyed Indians" were descended from Welsh forebears, "though I have no way of assessing their language."

Archeologists have drawn distinct parallels between the ruins of three pre-Columbian forts in Tennessee, Georgia, and Alabama, and the remains of ancient forts in Wales. But there is no proof that Prince Madoc and his band ever reached North America. At Mobile Bay, Alabama, however, a plaque today commemorates Madoc, "who landed on the shores of Mobile Bay in 1170 and left behind, with the Indians, the Welsh language."

Reclusive community

Throughout the United States there are numerous indications that North America might have been settled far earlier than is commonly believed. In Sneedville, Tennessee, a small, reclusive, close-knit group of people, known as Melungeons, are believed by many to have been in the vicinity long before the first Europeans and the Indians. According to eminent Orientalist, Dr. Cyrus H. Gordon,

Jews in India
Over 4,000 tribal people in the remote northeastern part of India have ardently embraced Judaism in the belief that they are descended from one of the Lost Tribes of Israel.

> When a supply ship from England reached the colony, it was deserted, with no sign that anything out of the ordinary had taken place.

the olive-skinned Melungeons may be descended from ancient Phoenicians who crossed the Atlantic centuries ago. Others scholars claim that they are the offspring of Portuguese colonists or sailors who settled in the region before the Revolutionary War. Still others point to them as proof that at least some of the Lost Tribes of Israel reached American shores. But no one can say for certain what actually happened.

Other tantalizing clues point to the possibility that even more "lost" peoples may have settled in North America. Near

Roanoke records
When Raleigh's first colonists arrived at Roanoke, they were accompanied by English artist John White, who drew this contemporary map of the island.

Los Lunas, New Mexico, is a rock on which is carved the Ten Commandments in what appears to be Phoenician. In the Pennsylvania State Museum at Harrisburg there are several large stones carved with what some experts believe to be Phoenician lettering. These were found nearby and the carvings have been judged to be over 2,000 years old. During the 1880's investigators unearthed stone tablets from prehistoric mounds near Davenport, Iowa. Among the inscriptions were characters that closely resembled Phoenician or Arabic.

Roanoke Island

While scholars debate the possibility that ancient Phoenicians might have been able to reach the shores of North America thousands of years ago, historians are still puzzled by a much more recent lost settlement.

In the 1600's, over a hundred settlers vanished without trace from a colony founded by Sir Walter Raleigh on Roanoke Island, off the coast of present-day North Carolina. When a supply ship from England reached the colony, the crew found it deserted, with no sign that anything out of the ordinary had taken place. On a post was carved the word *Croatoan*. The crew was baffled and for over a hundred years no one had a clue about the fate of the colonists.

In 1791, hunters chanced upon an unusual race of light-skinned American Indians, the Croatans, who lived about 200 miles west of Roanoke. The hunters were astonished to find that these mysterious Indians spoke English, lived in well-planned villages, and even used slaves to help them farm. A later census also showed that the Indians used many of the names of the first settlers. The Croatans also spoke of their blue-eyed ancestors who "talked in books."

It appears that the colonists may well have been abducted by their Croatan captors and forced to live out their lives among them as slaves. This may have been the fate of the other lost tribes of the world as well: they gradually disappeared as they intermarried with and adopted the lifestyle of their captors.

Modern ways
Clothed and fed, Ishi was photographed at the University of California in October 1911, two months after he stumbled into the 20th century.

STONE AGE MAN

In the early hours of August 29, 1911, the inhabitants of Oroville, California, discovered a near-naked, emaciated man in their midst. News of his discovery soon reached anthropologists who suggested that he might be a survivor of the Stone Age Yahi Indians. Leading an aboriginal existence, his ancestors had avoided discovery for centuries, and by the 20th century the tribe was thought to be extinct.

Ancient skills

Ishi (this means "a man" in Yahi) was moved to a residence at the Museum of Anthropology at the University of California at San Francisco, where he began to learn English and grew used to wearing clothes. He delighted visitors by demonstrating his prowess at such ancient skills as creating fire with a fire drill, stringing a bow, and making arrow heads.

In December 1914 Ishi fell ill. He developed a bad cough and went into hospital for tests. Having no immunity to 20th-century diseases, he had contracted tuberculosis, and died on March 25, 1916.

CHAPTER SIX

TOO GOOD TO BE TRUE

Man is a storyteller, and many of his best tales owe more to fiction than to fact. It appears that some of the more wondrous stories have been circulating for centuries. But the question remains: Why are people so willing to believe in them?

One winter night, as a trucker swung his 18-wheeler off the interstate highway onto the county road, his headlights picked out a lone female hitchhiker, no more than 18 years old. The driver knew that he was not allowed by his company to pick up hitchhikers, but he also knew that this was not the place and time for a teenage girl to be out alone.

Quickly, he brought the truck to a halt, and the girl climbed aboard. Over the roar of the engine, she told him that they were only about 10 miles from her destination, her hometown. In a few minutes, the lights of the town came into view. On Main Street the girl told the trucker to stop: she could walk home from there. She jumped down from the cab, and the two parted with a smile.

Lost property

At the next truck stop the driver discovered that the girl had left a book behind in his cab. He looked at the book and found her name and address on the flyleaf. The following week, when the trucker took the same route, the man called at the house to return the book. The girl's mother, who answered the door, listened to his story with tears in her eyes. The woman explained the cause of her distress: "My daughter used to hitch home from college every couple of weeks or so. Then, two years ago, just where you picked her up, she took her last ride — she was killed in an automobile accident."

Known as "The Vanishing Hitchhiker," this story, or variations on it, has been told for decades — in the United States, in Europe, in Japan and Korea. It is always passed on as something that has recently happened to a relative, or a friend of a relative, or a friend of a friend. The tale has been debunked scores of times on radio and TV and in print — yet many people still believe it and repeat it as a factual account.

Whale tumors

"The Vanishing Hitchhiker" is probably the best-known of the great number of widely circulated, deliciously apocryphal modern stories commonly termed urban legends, because they form a kind of modern counterpart to the legends of ancient times. Such stories are also known as FOAF ("friend of a friend")

tales, or "whale tumor" stories. The latter term derives from a story in a book by British folklorist Rodney Dale, *The Tumour in the Whale* (1978). Dale wrote that just after the Second World War, when people in Britain were encouraged to eat whale meat because it was cheap and plentiful, a grisly story circulated that someone had brought home a whale steak only to find that, when it was unwrapped, it was "gently throbbing" from a "live tumour" inside. This story was not true; for one thing, tumors do not "throb." The tale was probably inspired by an understandable suspicion of strange foods. And by the late 1940's this story had gone out of circulation.

"The Vanishing Hitchhiker," however, has had a much longer life, and a continuing one. Its origins can be traced far back in European folklore; and researchers have established that the same story was told in Illinois in 1876 and in Georgia in 1912. In these earlier versions the role of the trucker was played by a young horseman or buggy driver. The story seems in part to reflect drivers' curiosity about solitary people seen for an instant at the roadside. It is an extremely imaginative answer to the question: "What if I had stopped and picked up that person?"

Severed fingers

Many urban legends, however, conform more closely to the story about the whale tumor. They echo our fears of the unknown, or our constant concern with the dangers of urban life. Another classic highway story that has been circulating for decades tells of a motorist who slows down for a male hitchhiker and then, when the man starts to open the door, senses some menace about him and pulls sharply away. (In some versions the driver narrowly escapes being attacked and robbed by a gang of young thugs.) A dozen or so miles up the highway the

DETECTING A FOAF STORY

The just-plausible plots of urban legends often seem designed to demonstrate to the credulous listener that life can sometimes be as strange as fiction.

But typically such stories have certain telltale characteristics that point to legend rather than fact. One is that the account is never firsthand: it is always prefaced with such phrases as "This actually happened to a friend of mine..." or "My uncle told me that...."

Famous people

Another common characteristic is the involvement of a celebrity in the tale. A third is the failure of the story to stand up to truly logical examination. For example, in "The Vanishing Hitchhiker," it is not explained how the book the trucker discovers in his cab has material existence, while the young girl herself is a ghost.

A fourth giveaway sign of urban legends is their neat, well-crafted nature, which is so unlike the generally untidy and inconsistent accounts of real-life happenings, with their typical loose ends.

> "The Vanishing Hitchhiker" is probably the best-known of the great number of widely circulated, deliciously apocryphal modern stories commonly termed urban legends.

▸ PAGE 130

THE ANGELS OF MONS

Over the centuries, soldiers have told stories of the seemingly miraculous appearance of mysterious battlefield guardians — stories that often seem to be recycled from conflict to conflict.

THE BATTLE OF MONS, FLANDERS, in August 1914, was one of the first and most ferocious actions in the First World War. The British troops were greatly outnumbered by the Germans, yet held them off for days before finally withdrawing. Within a few months, word had spread in Britain that during the battle there had appeared a series of astonishing visions. These, it was claimed, had demoralized the attacking Germans with the result that the British escaped with far lighter casualties than could have been expected.

One British soldier was said to have told a Miss Marrable, an eminent clergyman's daughter, that "he and his company...heard the German cavalry tearing after them...they turned round to face the enemy, expecting instant death, when, to their wonder, they saw between them and the enemy a whole troop of angels, and the horses of the Germans turned round terrified...and tore away in all directions." Similar almost miraculous scenes were allegedly witnessed by thousands of other British troops.

"Complete fabrication"

Ultimately, the angels of Mons story became celebrated as a kind of "secret weapon" in discussions of the battle, and so, in a sense, the British were able to regard their military defeat as a psychological victory. However, Miss Marrable soon denounced as a "complete fabrication" the

Set to music
This sheet music is for a popular waltz that was inspired by the welcome news on the home front of divine intervention at Mons.

report that she had been told the story. It also turned out that one of the more celebrated witnesses to the angels' visitation, one Private Cleaver, of the 1st Cheshire Regiment, had actually been in England on leave at the time. In fact, not one genuine firsthand account of the "miracle" of Mons actually existed.

Archers' ghosts

The very first tale of paranormal events at Mons never pretended to be anything but pure fiction. It was a short story, "The Bowmen," a First World War tale by Arthur Machen, first published in the London *Evening News* on September 29, 1914, and later in book form. This story described an intervention at Mons, not of angels but of the ghosts of English archers who had overcome superior French forces at the battle of Agincourt in 1415. Machen denied that there was any truth in his tale. However, folklore soon translated the ghostly bowmen of fiction into a vision of angels at Mons. And from that time forward, the legend has been repeated time and time again.

Spiritual support
In his First World War story "The Bowmen," Arthur Machen invented a tale of the ghosts of 15th-century English archers coming to the assistance of 20th-century British troops.

Film mayhem
The "severed fingers" legend is echoed in the Australian film Mad Max, *in which it is a motorcycle hoodlum who loses his hand.*

THE MAXIMUM FORCE OF THE FUTURE

SAMUEL Z. ARKOFF Presents "MAD MAX"
Music by BRIAN MAY
Written by JAMES McCAUSLAND and GEORGE MILLER
Produced by BYRON KENNEDY Directed by GEORGE MILLER
with MEL GIBSON Color prints by MOVIELAB
R RESTRICTED
RELEASED BY AMERICAN INTERNATIONAL/A FILMWAYS CO.

late 1970's it had reached Australia, where a version of it was incorporated into the film *Mad Max* (1979). The story seems inextricably linked with the automobile age. But in his book *The Choking Doberman* (1984), American folklorist Jan Harold Brunvand reveals that the legend existed long before the age of the automobile.

A severed hand

In a 16th-century French work, *La Nouvelle Fabrique des Excellents Traits de Vérité* (1579), one story tells how a horseman is assailed by a would-be thief, who grabs his horse's bridle. The rider draws his sword and hacks at the villain, who lets go. The horseman forgets to tell his servant about the incident when he arrives home; the man is aghast to find a severed hand still holding the bridle.

The longevity of this highway legend is a telling indication that it reflects a perennial and widespread fear of lawless and desperate characters — and a fear, too, of being drawn into their world of terrifying violence by resisting them.

The mad axman

Another such tale with a gruesome and ironic twist concerns two coeds who stay alone in their college dorm during one Christmas vacation. They know that a "mad axman" has escaped from the local penitentiary, so they take extra care over security, until one night one of the girls decides to use the library. Some time after she has gone, her roommate hears a

driver stops for gas, but when the pump attendant goes to fill up the tank, he halts in speechless horror and points at the passenger door. The driver gets out to look. Caught in the door handle are three severed fingers. The "severed fingers" tale was popular in Britain in the early 1960's, and by the

weird shuffling sound, like that of feet dragging down the hall toward the room. Afraid, she double-locks and barricades the door. The shuffling sound stops, and then a horrible scratching at the door begins. It goes on for hours while the terrified girl cowers inside. Then it ceases. Finally, the exhausted girl falls asleep.

When the girl wakes up the following morning, she wonders what has happened to her friend

Whale tumor victim
Public figures often appear in FOAF stories. Several such tales have circulated about Michael Jackson.

and rushes to open the door. There, lying dead on the floor outside, is her friend, with a hatchet buried in her back. The scratching sound that had so terrified her had not been that of the killer but of his victim, the girl's dying roommate, desperately trying to get help.

Barefoot Beatle

Urban legends about celebrities are also common. In the late 1960's, for example, a story that Beatle Paul McCartney was dead circulated among teenagers. Much evidence was cited in support of this story, including details of the picture on the cover of the group's album *Abbey Road* (released in 1969). It was believed to be of great significance, for example, that Paul McCartney was the only barefoot member of the group on the album cover. The story petered out when a public appearance established that the most popular Beatle was alive and well.

> According to another Jackson legend, the rock music channel MTV announced that the first seven digits of the catalog code on Jackson's album *Thriller* were the same as his private, unlisted telephone number.

Another much repeated story is that on *The Tonight Show*, host Johnny Carson made various sexually suggestive remarks to famous guests, such as Zsa Zsa Gabor. But no one who tells this story was ever actually watching and heard these remarks when they were made — the witness is always a friend or a relative of the storyteller.

Fatal accident

Another superstar, the reclusive Michael Jackson, has had many FOAF tales of this kind told about him. One claims that the then 12-year-old dancer Alfonso Ribeiro, who appeared with Jackson on a TV commercial, later died while break

dancing. However, this did not happen.

According to another Jackson legend, the rock music channel MTV announced that the first seven digits of the catalog code on Jackson's album *Thriller* (released in 1982) were the same as his private, unlisted telephone number. As a result, the staff of a hair studio in Bellevue, Washington, whose number actually did coincide with the digits, were at one time receiving up to 50 calls a day from insistent fans demanding to speak to their hero.

Cadillac reward

Other urban legends have sprung up around celebrities, or their relatives, that show these people to be exceptionally kind and generous. Such a story has been told separately about the wives of boxer Leon Spinks and singer Nat "King" Cole. According to this particular legend, the woman's automobile breaks down, a motorist who happens to be passing fixes the engine, and as a reward, the grateful celebrity's wife gives her rescuer a new Cadillac.

Generous stars

Another legend of this kind concerns Burt Reynolds. Allegedly, the star announced — on Johnny Carson's *The Tonight Show* once again — that he had just won a million-dollar libel suit against the *National Enquirer* and that, to share his good fortune, anyone could, for a limited period, charge their telephone calls to his telephone credit card number, which he duly supplied. The same tale has been told, at various times, about Steve McQueen, Robert Redford, and Paul Newman. Like Reynolds, however, none of them has ever filed a lawsuit against the *National Enquirer*.

One of the most profound and lasting mysteries about FOAF's is who invents them. As with most jokes, and certain manufactured terms, such as the ever-useful "gizmo," it is all but impossible to track them back to their origins.

Free calls
Film superstar Burt Reynolds is the subject of a modern legend. He is falsely reputed to have told viewers of The Tonight Show *that they could charge their telephone calls to his credit card number.*

A QUESTION OF REALITY

Where do strange stories come from? They are not always invented. Sometimes they seem to arise from a highly subjective interpretation of quite ordinary events.

IT WAS AUGUST 21, 1915. The British and colonial troops fighting the Turks at Gallipoli were meeting determined resistance in Suvla Bay. According to three eyewitnesses, all of whom were soldiers from New Zealand, a solitary group of clouds remained oddly motionless in the sky during this breezy day. One cloud, an "absolutely dense, solid-looking structure," seemed to be resting on the ground near Hill 60. Then, claimed the witnesses, who were positioned in a neighboring sector, the First-Fourth Norfolks were observed marching up a sunken road or creek toward Hill 60. "However, when they arrived at this cloud, they marched straight into it...but no one ever came out to deploy and fight at Hill 60. About an hour later, after the last of the file had disappeared into it, this cloud very unobtrusively lifted off the ground and...rose slowly until it joined...other similar clouds which...all moved away northwards.... Those who observed this incident can vouch for the fact that Turkey never captured that regiment."

False account

This account was first published in 1965, the 50th anniversary of the campaign. It has been repeated often in books about unexplained mysteries. Yet there are plenty of errors in the story, as British researcher Paul Begg reported in his book *Into Thin Air* (1979). For example, the First-Fourth Norfolks is not a regiment in its own right but the First Battalion of the Fourth Norfolk Regiment. Also, this battalion did not disappear, but went on to fight for the rest of the campaign.

However, on August 12, not August 21, one British force did vanish — it consisted of more than 250 men of the First Battalion of the Fifth Norfolk Regiment. However, this was at night, and the men were nowhere near Hill 60.

The First-Fifth Norfolks
These are the men of the First Battalion of the Fifth Norfolk Regiment who disappeared at Gallipoli on August 12, 1915. The battalion was made up mainly of men recruited from the British Royal Family's estate at Sandringham in Norfolk. Some of the men were known personally to the Royal Family, as they worked either on the estate or in the royal household.

The Gallipoli campaign was one of the most disastrously managed of the war; conditions were appalling, the situation constantly confused, and of the 34,000 British and allied troops who died, some 27,000 have no known grave. So it is hardly surprising that 250 men might have vanished at night in a single incident or that a legendary story about vanishing troops might have sprung into existence.

Outlandish tale

But why should three New Zealand veterans concoct the outlandish tale that was published? Perhaps the real answer is that these men did not do so deliberately but may unwittingly have fused together several separate elements: their own 50-year-old memories of August

> The case of the First-Fourth Norfolks shows how arbitrary a version of reality we are sometimes led to accept.

1915, and the two accounts of the campaign that appeared in an official publication they may very well have read, *The Final Report of the Dardanelles Commission* (1965). The first account describes the disappearance of the First-Fifth Norfolks on August 12. The second account describes the unusual weather conditions of August 21, when Australian and New Zealand forces attacked Hill 60: "By some freak of nature Suvla Bay and Plain were wrapped in a strange mist on the afternoon of 21 August." Therefore a combination of creative, or possibly errant, memory and an innate human tendency to mythologize may account for the creation of the First-Fourth Norfolks

CONFUSION OF COLORS

Seeing is not always believing. The following experiment shows how easily our senses can be led astray.

SOMETIMES THE CONTEXT in which we see things may cause us to identify them falsely. In the chart below the names of various colors are each printed in a color different from that which the letters designate. Ask a friend to name, quickly, the color in which each name is written here. Most people stumble over this task, giving the printed name rather than the actual color it is written in.

Psychologists call this phenomenon the "Stroop effect." It occurs because the dominant verbal part of the mind finds it extremely difficult to yield to that part of the mind that perceives color.

Reverse hills

Motorists who are convinced of the existence of "reverse hills" (those on which, they claim, freewheeling automobiles seem to roll uphill) are also the victims of conflicting mental interpretations. In this case, they allow certain misleading optical clues to override the evidence of their other senses.

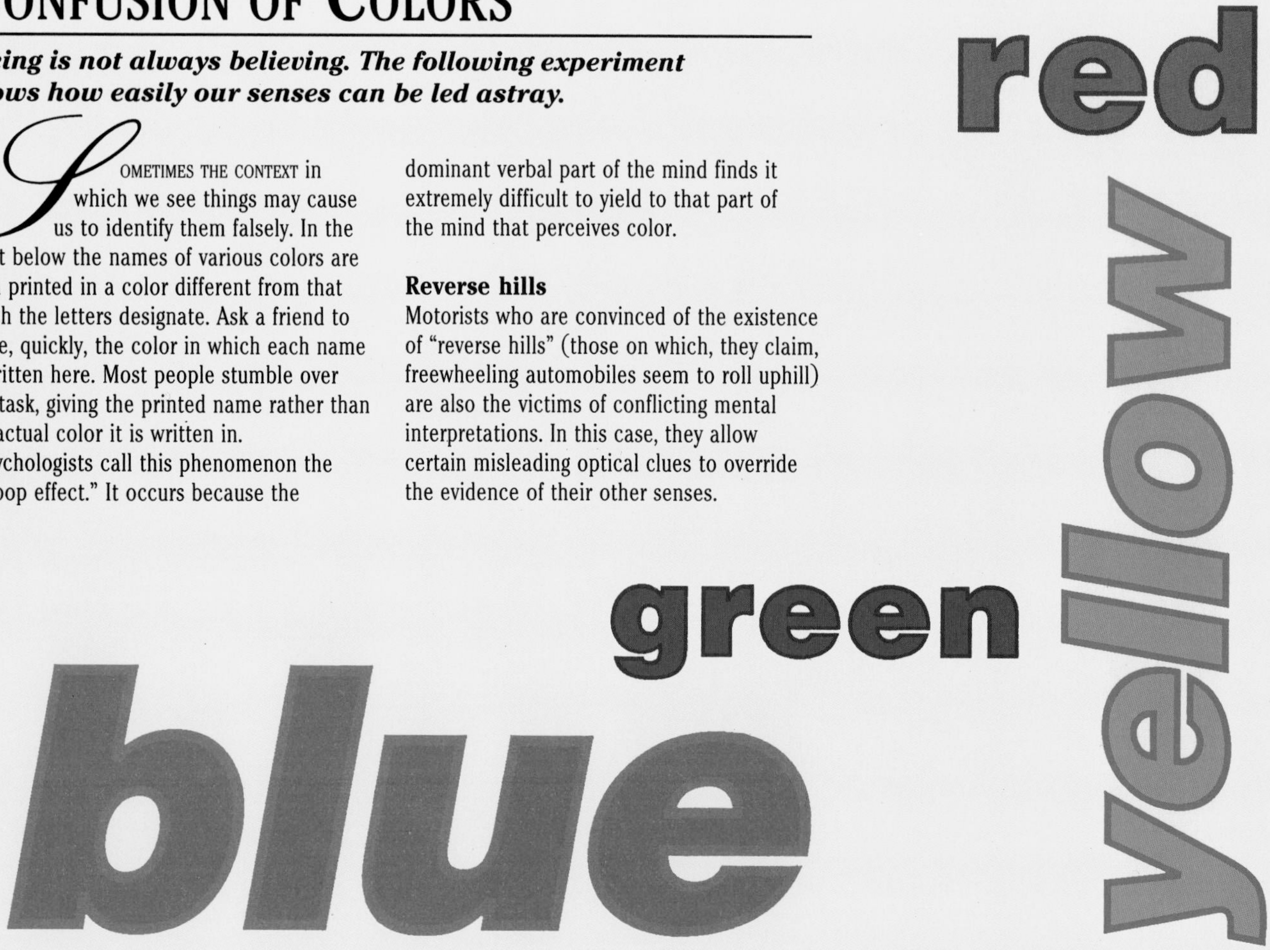

red

yellow

green

blue

legend. Those who told the story did so in good faith. To the veterans, the illusion was as utterly convincing as the actual facts would have been.

Psychological test

The case of the First-Fourth Norfolks shows how arbitrary a version of reality we are sometimes led to accept. Far from viewing the world objectively, as most of us like to think we do, we see it, rather, filtered through our expectations and preconceptions. This hypothesis has been shown in the following test devised by experimental psychologists. By bringing the two digits of the number 13 close together, it can be made to look like the letter B — and it has been discovered that, printed thus and surrounded by other letters, the figure will indeed be identified as a B. Nevertheless, if it is surrounded by numerals, it will be interpreted as 13. In other words, we tend to find exactly what we are looking for. Those with a strong belief in the paranormal will see its workings in phenomena that rationalists will regard as having a logical explanation. Thus it has been pointed out that religious miracles are almost exclusively seen by religious people.

It may be for this reason that some people have seen letters or numbers in the iris of an eye, faces in the stains on a plaster wall, and other strange images where they have no reason to be.

Unconscious dupes

This example should provide some indication of how easy it is to be misled on occasion by the sensory input on which we depend for our (generally very reliable) interpretation of the world around us. Some of those who witness certain miracles and wonders may be well-meaning but the

A casebook of mistaken identity

Murderous cleric
The story of the phantom vicar of Wapping, London, was told in the British television program A Leap in the Dark.

unconscious dupes of such sensory illusions. In some brain diseases, inner visual interpretation may be sufficiently disturbed to produce truly extraordinary behavior. For example, in his best-selling book *The Man Who Mistook His Wife for a Hat* (1986), the eminent British-born neurologist Oliver Sacks, of the Albert Einstein College of Medicine, New York, tells the strange and tragic story of a man who literally made the mistake of the title, reaching out for his wife's head and trying to lift it off and put it on his own.

Hostile scientists

Yet the scientific establishment has traditionally remained extremely hostile to claimed evidence for paranormal phenomena. An example of this occurred in 1981 following a fascinating announcement by French statisticians Michel and Françoise Gauquelin that, in studying the astrological birth charts of more than 2,000 champion athletes, they had discovered that the planet Mars was either rising or transiting at the time of birth of 22 percent of them. (Pure chance would have dictated that only 17 percent of the athletes should have been born at such a

PHANTOM VICAR

In the summer of 1970 English journalist Frank Smyth wrote an article for the magazine *Man, Myth and Magic*, in which he invented a 19th-century clergyman and his ghost. Smyth's imaginary vicar, the owner of a seamen's rooming house at Ratcliffe Wharf in Wapping, a seedy area of London's dockland, robbed and killed his boarders and threw their bodies in the Thames. Since the cleric's death, Smyth wrote, his remorseful spirit had haunted the district.

Many "witnesses"

Over the following year eight books on ghosts included the phantom vicar. When Smyth appeared on a BBC2 television program in 1973 about the hoax, *A Leap in the Dark*, a number of interviewees claimed that they had actually seen the clergyman's ghost. The novelist Jilly Cooper told of a police superintendent who had recounted that as a young man — decades before Smyth had invented the ghost — he had avoided Ratcliffe Wharf for fear of meeting it. And other "witnesses" wrote in to the TV station, refusing to accept Smyth's story of a hoax and claiming to have seen the apparition.

Before Smyth's story there were no written or oral accounts of a phantom vicar haunting Wapping or any other part of London's dockland. The episode demonstrates the remarkable suggestibility of many people.

> "Man tries to make for himself...a simplified and intelligible picture of the world.... He makes this cosmos and its construction the pivot of his emotional life in order to find in this way the peace and serenity which he cannot find in the narrow whirlpool of personal experience."
>
> **Albert Einstein**

time.) A scientific body, the Committee for the Scientific Investigation of Claims for the Paranormal (CSICOP), reacted at once to the findings by attempting to discredit the Gauquelins' methods of statistical analysis, as did the Comité Para, the Belgian equivalent of CSICOP. The Gauquelins fought back, demolishing all their critics' many objections and thereby proving themselves the better statisticians. CSICOP and the Comité Para then checked the data again and reached the same findings as the Gauquelins.

Excellence in sport

Both CSICOP and the Comité Para raised criticisms about the Gauquelins' methods. But they had not used their research to support astrology. They had not claimed, as an astrologer would have done, that the position of Mars at the time of the athletes' birth caused them to excel in sport. They had said that there was a correlation between two sets of data, and that it was an irregular one. This was not a big claim, but it had disturbed the scientists.

The reason that the classic scientific mind reacts with such hostility to any claimed evidence of the paranormal may be rooted in the basic assumptions that underlie scientific method. Scientists carry out experiments to test specific hypotheses. Hypotheses that appear to pass such tests are, in some cases, put forward as theories or facts, some of which scientists may one day refer to as laws of nature. Those hypotheses that fail experimentally are discarded. Behind this method lies the scientist's belief that whatever he or she observes during experiments is the sole form of reality.

In his famous book *Zen and the Art of Motorcycle Maintenance* (1975) — an intriguing attempt to find a balance between objective and subjective ways of treating experience — Robert M. Pirsig

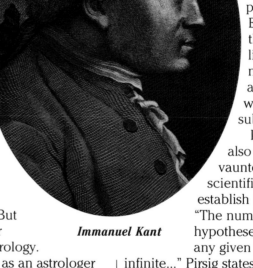

Immanuel Kant

quotes Albert Einstein's explanation of why a definite belief in some form of ultimate objective reality is so comforting: "Man tries to make for himself...a simplified and intelligible picture of the world....He makes this cosmos and its construction the pivot of his emotional life in order to find in this way the peace and serenity which he cannot find in the narrow whirlpool of personal experience." Einstein saw that there was no clear link between the nature of the cosmos and the world that we all experience subjectively.

In his book Pirsig also questions the much-vaunted ability of the scientific method to establish certain truths. "The number of rational hypotheses that can explain any given phenomenon is infinite..." Pirsig states. "If the purpose of scientific method is to select from among a multitude of hypotheses...then it is clear that all hypotheses can never be tested. If all hypotheses cannot be tested, then the entire scientific method falls short of establishing proven knowledge."

Parallel cosmos

Important questions arise from this argument. Could the untested (perhaps untestable) hypotheses account for what lies behind "paranormal" happenings? Could these paranormal phenomena exist in a cosmos parallel to that of scientific laws? Are both equally real?

Thus it is at least conceivable that the perception of paranormal phenomena — like telepathy and repeated sightings of UFO's — is a sign of mental powers that, as the German philosopher Immanuel Kant (1724–1804) remarked of time, we recognize but cannot directly examine. Another possibility is that paranormal phenomena may simply be creations of the human mind: frogs fall from the sky because, consciously or unconsciously, the witness has wanted it to happen.

MANY MEANINGS

Ancient inscriptions in unknown languages provide an opportunity for those deciphering them to read into them whatever meaning they may choose.

SIX-AND-A-HALF MILES SOUTH of Taunton, Massachusetts, at Dighton, on the east side of the River Taunton, lies what is known as the Dighton Writing Rock, a five-foot-high boulder of graywacke, a form of sandstone. Carved into its face, which is about 11 feet 6 inches across and fronts the river, are a number of mysterious hieroglyphics.

Since the rock's discovery by New England colonists in 1680, various interpretations have been placed upon these complex and enigmatic markings. In 1893, for example, G. Mallery, in his book *Picture-*

Dighton Writing Rock
The first "faithful and accurate representation" of the markings on the rock is this drawing by Dr. Danforth in 1680.

'melek' (king). Another scholar triumphantly established the characters to be Scythian, and still another identified them as Phoenician."

American Indian burial ground

In 1838 another mysteriously marked stone was discovered: an inscribed tablet of stone excavated from an American Indian burial ground at Grave Creek, Ohio. In his book *Megalithomania* (1982), the British historian John Michell enumerates the various interpretations that scholars down the years have placed upon the characters. Some have interpreted them as being Etruscan, ancient Greek, Runic, ancient Gaelic, old Erse, Phoenician, old British Celtiberic, and Canaanite. Translations of the inscription are equally diverse. They include:

"What thou sayest, thou dost impose it, thou shinest in thy impetuous elan and rapid chamois."

"The Chief of Emigration who reached these places (or this island) has fixed these statutes forever."

"The grave of one who was assassinated here. May God to avenge him strike his murderer, cutting off the hand of his existence."

Doubtless, there will be further interpretations of the hieroglyphics. It may be, however, that the inscriptions on both of these rocks mean, in fact, nothing at all.

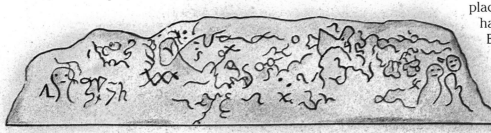

Greenwood's drawing
This drawing of the Dighton Rock inscription by Dr. Isaac Greenwood, exhibited to the Society of Antiquarians in London in 1730, is considerably different from Dr. Danforth's.

Writing of the American Indians, cited an account that dismissed claims that the inscription resembled ancient Roman letters and figures and that confidently asserted that the inscription "is of purely Indian origin, and is executed in the peculiar symbolic character of the Kekeewin." Mallery continued: "The...characters were translated by a Scandinavian antiquary as an account of the party of Thorfinn, the Hopeful. A distinguished Orientalist made out clearly the word

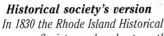

Historical society's version
In 1830 the Rhode Island Historical Society produced yet another representation of the hieroglyphics.

THE COSMIC JOKER

We have all felt manipulated by events that seem beyond our control. At other times bizarre coincidences occur that astound us. Charles Fort, collector of such anomalies, saw these experiences as being the work of the Cosmic Joker.

IN OCTOBER 1987, on the Caprivi Strip bank of the Zambezi River in Namibia, postmaster Kobus Slabbert warned the local children that they should not swim in the river because of the danger from man-eating crocodiles. As he was speaking, a crocodile emerged from the river behind him, caught him by the leg, and pulled him into the water. His body was never recovered.

On another occasion in Prague, Czechoslovakia, a woman was reported to be so depressed by her husband's constant infidelity that, intending to commit suicide, she threw herself from a third-floor window. Her fall was broken by a hapless man who was just entering the building. The man was killed. And her victim was none other than her errant husband.

In his writings the phenomenologist Charles Fort occasionally speculated that such episodes were the work of a power that he termed the Cosmic Joker. The idea of a transcendent trickster, toying with the fate of humanity, is found in many cultures. In ancient Greek legend, for example, the role was performed by a gaggle of malicious hags known as the Furies, and in the Christian tradition the function is attributed to the devil.

The idea of a transcendent trickster, toying with the fate of humanity, is found in many cultures.

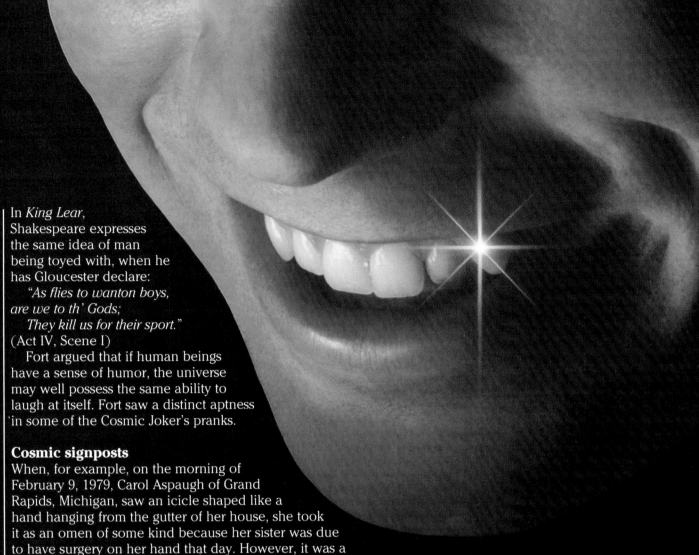

In *King Lear*, Shakespeare expresses the same idea of man being toyed with, when he has Gloucester declare:

"*As flies to wanton boys, are we to th' Gods; They kill us for their sport.*" (Act IV, Scene I)

Fort argued that if human beings have a sense of humor, the universe may well possess the same ability to laugh at itself. Fort saw a distinct aptness in some of the Cosmic Joker's pranks.

Cosmic signposts

When, for example, on the morning of February 9, 1979, Carol Aspaugh of Grand Rapids, Michigan, saw an icicle shaped like a hand hanging from the gutter of her house, she took it as an omen of some kind because her sister was due to have surgery on her hand that day. However, it was a more direct cosmic signpost than anyone could have guessed. The surgeons were unable to operate that day because Carol Aspaugh's sister had had an accident: a falling icicle had injured her hand.

The Austrian psychologist Sigmund Freud believed that jokes are the result of allowing chaos to intervene briefly in the conventional order of our lives through some grotesque or unexpected juxtapositions of ideas, images, actions, or words. If we follow Fort's hypothesis that the universe, too, possesses a sense of humor, then strange phenomena may be the way it expresses it — that is, by breaking its own rules from time to time.

Fort believed that people, animals, and objects could be carried by unidentifiable means to other locations, a kind of instant transportation that, in *The Book of the Damned* (1919), he termed teleportation. Fort believed teleportation to be linked to the awesome natural force that shifted vast amounts of matter in the formation of the planet. He saw it as the explanation for various other phenomena such as falls of fishes from the sky

"As flies to wanton boys, are we to th' Gods; They kill us for their sport."

William Shakespeare

that would otherwise be inexplicable. Fort could also explain other anomalous phenomena — from, for example, out-of-place animals to simulacra — by ascribing them to the Cosmic Joker.

Intolerant attitude

In Western society, skeptics are not amused by reports of weird phenomena, and people who have had inexplicable experiences or have shown an interest in the paranormal have often been ostracized. Such deep, and fiercely defended, divisions of humankind into believers and non-believers is unfortunate and might be regarded by those with a Fortean cast of mind as tragic. This is not because one side or the other is right, but because both are, in a sense, wrong in clinging to a single rigid and therefore limited view of the miraculous world in which we live. Instead of celebrating and laughing at the richness of life's strangeness, and its often apparent absurdity, we have perhaps lost some of our capacity to cherish its enduring wonder.

TEXT SOURCES

The following is a list of sources for the bizarre phenomena described in this book. Page numbers and subject areas are in **bold** type.

Chapter 1: Exceptional Energies
22–25: Reeser: F. Hutching: *World Atlas of Mysteries* (Pan Books, UK, 1979); J. Michell & R. Rickard: *Phenomena* (Thames & Hudson, UK, 1977); C. & J. Bord: *Modern Mysteries of the World* (Grafton, UK, 1989); *Mysteries of the Unexplained* (Reader's Digest, 1982); V. H. Gaddis: *Mysterious Fires & Lights* (Dell, 1967). **25: Burning hands**: J. Nickell & J. F. Fisher: *Secrets of the Supernatural* (Prometheus Books, 1988); **Angel**: *Fate* (Sep. 1982); **Winchester**: *Fortean Times*, UK (Spring 1983); **Hayes**: *Evening Standard*, London (May 31, 1985); *National Enquirer* (Jul. 23, 1985). **26: Parisian woman**: T. Bartholin: *Acta* (Copenhagen, 1673); **Millet**: J. Dupont: *De Incendiis Corporis Humani Spontaneis* (Leyden, 1763); **19th-cent. theory**: *Secrets of the Supernatural*; **Modern explanations**: I. T. Sanderson: *Investigating the Unexplained* (Prentice-Hall, 1972); *Applied Trophology* (Dec. 1957). **26–27: Gearhart theory**: *Pursuit* (1975). **27: Static electricity**: *Mysteries of the Unexplained*; **Human candle theory**: *Q.E.D.*, UK TV program (Apr. 26, 1989). **28: Supino**: *Sunday Mirror*, London (Aug. 21, 1983); **Sascha K.**: *Izvestia*, U.S.S.R. (Apr. 11, May 18 & 23, 1987); *Stars & Stripes* (Apr. 14, 1987). **28–29: Compton**: C. Compton & G. Cole: *Superstition* (Ebury Press, UK, 1990). **29: Lahore family**: *Daily Mirror*, London (Aug. 16, 1979); *Daily Mail*, London (Sep. 11, 1979); *Superstition*; **Wharncliffe**: *Houston Post* (Jun. 16, 1983); *Columbus Dispatch* (Jul. 24, 1983). **30: Brownell & O'Grady**: *Duluth Tribune & Herald*, Minnesota (Mar. 1, 1985). **30–32: Marsden**: *Portsmouth News*, UK (Jan. 25, 1989). **32: Orkney outing**: *Scotsman*, UK (Jul. 18–19, 1990). **33: Vanishing ship**: C. Berlitz & W. Moore: *The Philadelphia Experiment* (Grosset & Dunlap, 1979); D. Clark: *Vanished!* (Michael O'Mara Books, 1990); **Magnetic cloud**: *Fortean Times*, UK (Spring 1987). **34: French meteorite**: *Phenomena*; **Frogs & fish**: Pliny the Elder: *Historia naturalis*. **34-36: Cardan's theory**: J. Cardan: *De subtilate* (1550). **36: Blood from sky**: *Report of the Brit. Assn. for the Advancement of Science* (1860); **Eels, ants, worms**: C. Fort: *The Book of the Damned* (Boni & Liveright, 1919); **Geese**: *Journal of Meteorology*, UK (1979). **36–37: Walker**: *Camden News*, Arkansas (Jan. 2, 1973). **37: Ice falls**: *L'Atmosphère*, France (1888); *The Times*, London (Aug. 14, 1849); *New York Post* (Apr. 26, 1989); *Le Quotidien*, Réunion Island (Aug. 16, 1989); *Sydsvenska Dagbladet*, Sweden (Apr. 28, 1990); *Weatherwise* (Jun., 1960); **Albino frogs**: *Daily Mirror & Daily Star*, London (Oct. 24, 1987); *Stroud Observer*, UK (Nov. 12, 1987); *The Times*, London (Feb. 8, 1988); *Natural World*, UK (Spring 1988); *Gloucester Citizen*, UK (Jun. 13, 1988). **38: Shad shower**: *Boston Globe* (Aug. 24, 1984); **Flying starfish**:

Minneapolis Star & Tribune (Apr. 24, 1985); **Dead sardines**: *Brisbane Courier Mail*, Aus. (Feb. 7, 1989); *Australian Post* (Mar. 18, 1989); **Heavenly windfall**: *Daily Mail*, London (Nov. 10, 1984). **39: Stanford apports**: *Encyclopedia of Occultism & Parapsychology* (Gale Research, 1988); *Journal of the Soc. for Psychical Research*, UK (Apr. 1987); **Sai Baba**: E. Haraldsson: "*Miracles Are My Visiting Cards*" (Century, UK, 1987). **41: Rhine experiments**: *Journal of Parapsychology* (1943); **Kulagina**: D. Scott Rogo: *The Poltergeist Experience* (Aquarian Press, 1979). **41–43: SORRAT**: *Research in Parapsychology* (1982); *The Unexplained* (Orbis, UK, 1981); info supplied by J. T. Richards. **43: Wormholes**: *Reader's Digest Library of Modern Knowledge* (UK, 1978); J. Michell & R. Rickard: *Living Wonders* (Thames & Hudson, UK, 1982). **44: Mrs. Anna**: E. U. Condon: *Scientific Study of UFO's* (Bantam, 1969); **Gwinnet**: G. T. Meaden: *The Circles Effect & its Mysteries* (Artetech Publishing, UK, 1990) & *Circles from the Sky* (Souvenir Press, UK, 1991); **Nippon Radio**: Y. H. Ohtsuki: *Journal of Meteorology*, UK (Vol. 17, No. 168, 1992); **Sacred circles**: G. T. Meaden: *The Stonehenge Solution* (Souvenir Press, UK, 1992). **46: Mowing devil**: ed. R. Noyes: *The Crop Circle Enigma* (Gateway Books, UK, 1990); **Wondrous markings**: *Fortean Times*, UK (Winter 1989); **Paranormal phenomena**: *The Crop Circle Enigma*; **Alien marks**: *Sunday Correspondent*, London (Jul. 29, 1990). **47: Bower & Chorley**: *Today*, London (Sept. 9, 10, & 11, 1991). **48: Barnes**: *Circles from the Sky*; **Reid**: *Circles from the Sky*; **Kikuchi**: info supplied by G. T. Meaden; **Swastika pattern**: *The Crop Circle Enigma*; **Graham**: *The Circles Effect & its Mysteries*; **Hawking**: *Cambridgeshire Times*, UK (Aug. 1991). **49: Tomlinsons**: *Mail on Sunday*, London (Aug. 25, 1991). **50–51: Battles**: *Journal of the Soc. for Psychical Research*, UK (May–Jun. 1952). **52: Energy**: *The Illustrated Thesaurus of Physics* (Cambridge Univ. Press, UK, 1985). **53: Kelvin & Rutherford**: C. Morgan & D. Langford; *Facts & Fallacies* (Webb & Bower, UK, 1981); **Seven dimensions**: *Facts & Fallacies* (Reader's Digest, 1988).

Chapter 2: Mysterious Animals
54–56: Thompson: L. Coleman: *Curious Encounters* (Faber & Faber, 1985). **56: Giant lizards**: L. Coleman: *Mysterious America* (Faber & Faber, 1983). **56–58: NY sewers**: *Mysterious America*; T. Pynchon: *V* (Bantam Books, 1963). **57: Eastern cougar**: *Mysterious America*, info supplied by P. Sieveking; K. P. N. Shuker: *Mystery Cats of the World* (Robert Hale, UK, 1989); **Gottschalk**: *Washington Post* (Oct. 8, 1990); *New York Times* (Jan. 4, 1991); **Attacks on humans**: *Daily Telegraph*, London (Apr. 26, 1991). **58: Alligators**: *Mysterious America*; **Mrs. Glover**: R. H. Gollmar: *My Father Owned a Circus* (Caxton,

1965). **59: Staub**: *Mysterious America*; **Chicago cops**: *Fate*, Vol. 31, No. 4, 1978; **U.S. kangaroos**: *San Francisco Chronicle* (Oct. 19, 1974); **Sutts**: *Mysterious America*; **Apelike creature**: *National Insider* (Nov. 2, 1975); **English crocodile**: *Gentleman's Magazine*, UK (Aug. 1866). **60: Te Kuiti**: *New Zealand Herald* (Dec. 9, 1982); **Early records**: William of Newburgh: *Historia rerum Anglicarum* (England, 1196–98); A. Paré: *Exemples des monstres que se font par pourriture et corrupt* (France, 1575); R. Plot: *Natural History of Staffordshire* (England, 1686); *Mémoires of the French Academy of Sciences, 1719 & 1731;* **Clarke**: W. Howitt: *History of the Supernatural* (UK, 1863); **Nat. hist.**: Mrs. Loudon: *Entertaining Naturalist* (UK, 1850). **60–61: Owen**: Lynn Barber: *The Heyday of Natural History, 1820–1870* (Jonathan Cape, UK, 1980). **61: Capt. Buckland**: J. Michell & R. Rickard: *Phenomena* (Thames & Hudson, UK, 1977); **Rev. Taylor**: *Hartlepool Mail*, UK (Aug. 1, 1980); **Siberian tundra**: *The Times*, London (Jun. 26, 1987); **Dr. Buckland**: Philip Gosse: *The Romance of Natural History* (UK, 1860); **Wood**: J. Michell & R. Rickard: *Living Wonders* (Thames &.Hudson, UK, 1982); **Mrs. Culp**: *Fate* (Jan. 1954); **Hardman**: *Strand Magazine*, UK (1901); **Liesky**: R. Bakewell: *Bakewell's Introduction to Geology* (UK, 1813). **62: Rennie**: *Phenomena*; **Minster**: J. F. McCloy & R. Miller, Jr.: *The Jersey Devil* (Middle Atlantic Press, 1976); *Mysterious America*. **63: Theories**: *The Unexplained* (Orbis, UK, 1981); **Topsham**: C. Fort: *The Book of the Damned* (Boni & Liveright, 1919); **Galicia**: *Illustrated London News* (Mar. 14, 1855); **Black dogs**: *Phenomena*. **64: Frith**: *Phenomena*; **Medieval legend**: William of Malmesbury: *Gesta pontificum* (England, 1123). **66: Gerbils, toads, porcupines**: C. Lever: *The Naturalised Animals of the British Isles* (Granada, UK, 1977); **Beast of Bath**: *Sunday Express*, London (Jul. 29, 1979); *Bath Evening Chronicle*, UK (Jul. 30, Aug. 1, 7, & 10, 1979); **Bear**: J. & C. Bord: *Modern Mysteries of Britain* (Grafton Books, UK, 1987); **Wild boars**: *Aldershot News*, UK (Aug. 8 & 15, 1972); *Daily Express*, London (Jul. 7, 1975); **Baboon**: *Evening News*, London (Sep. 30, 1976); **Monkeys**: *Leicester Mercury*, UK (Sep. 7 & 8, 1979); *Daily Express*, London (Sep. 10, 1979); **Other colonists**: *The Naturalised Animals of the British Isles. Journal of the Camborne-Redruth Nat. Hist. Soc.* (UK, 1974). **67: Brazil, dolphins**: *Los Angeles Times* (Mar. 26, 1990); **Experiments**: P. G. H. Evans: *The Nat. Hist. of Whales & Dolphins* (Christopher Helm, UK, 1987); **Elberfeld horses**: M. Maeterlinck: *The Unknown Guest* (UK, 1914); **Sparkie**: *Guinness Book of Records* (1958); **Alex**: J. Sparks & T. Soper: *Parrots: A Nat. Hist.* (Facts on File, UK, 1990); **Hoover**: *Lincoln Journal*, Nebraska (Jun. 25, 1981). **68–69: Minnesota apeman**: *The Unexplained* (1982); *Living Wonders*. **69: Talking mongoose**: H. Price & R. S. Lambert: *The Haunting of Cashen's Gap* (Methuen, UK, 1936); *The Unexplained* (1982). **70: Col. Fordyce**: *Scots Magazine*, UK (Jun. 1990). **70–71: Nessie** theories: *Nat. Hist.*, UK (1934); R. T. Gould: *The Loch Ness Monster & Others* (Geoffrey Bles, UK, 1934); *New Scientist*, UK (1960 & 1979); *Mysterious Creatures* (Time-Life Books, 1988). **71: Champ**: M. Meurger & C. Gagnon: *Lake Monster Traditions* (Fortean Tomes, UK, 1988); R. Mackal: *Searching for Hidden Animals* (Doubleday, 1980).

Chapter 3: Expect the Unexpected
72: Springer/Lewis: *The Unexplained* (Orbis, UK, 1982); P. Watson: *Twins* (Hutchinson, UK, 1981). **74: Twins**: *Encyclopaedia Britannica* (1983). **75: Crominski**: *Daily Express*, London (Feb. 7, 1978); **Johnson/Harcourt**: *The Leicester Mercury,* UK, (Oct. 28, 1988); **Standon/Garber**: *Herald*, Circleville, Ohio (May 11, 1988); **Barnoldswick**: *Daily Express*, London (Apr. 14, 1988). **76: Hewitt/Fletoridis**: *Daily Express*, London (Jul. 25, 1987); **Harris**: *Associated Press* (Mar. 6, 1991); **Ackroyd**: *Daily Star*, London (Jun. 27, 1991); **Hume**: *Enterprise*, Texas (Jan. 31, 1981); **Hawsawi**: *Saudi Gazette*, (Apr. 15, 1986). **77: Milai, twin study**: *Observer*, London (Dec.15, 1991). **78: Jay**: I. Stevenson: *Unlearned Language* (Univ. Press of Virginia, 1984). **79: Neuman**: *Encyclopedia of Occultism & Parapsychology* (Gale Research, 1988); **Frenchwoman, cryptomnesia**: *Harper's Encyclopedia of Mystical & Paranormal Experiences* (Harper, 1991); **Gobbledegook**: W. J. Samarin: *Tongues of Men & Angels* (Macmillan, 1972); **Huddar**: *Unlearned Language*. **80: Gregory of Tours**: P. Devereux: *Earth Lights Revelation* (Blandford, UK, 1990). **81: Great Airship**: *Earth Lights Revelation* ; **Mysterious Rockets**: *The Age of the UFO* (Orbis, UK, 1984); **Adamski**: D. Leslie & G. Adamski: *Flying Saucers Have Landed* (Werner Laurie, London, 1953); **Meier**: L. J. Elders & T. K. Welch: *UFO...Contact from the Pleiades* (Genesis III Pub., 1980). **82: Arnold G.**: *Sunday Express*, London (Oct. 21, 1984); **Watling**: *Evening Standard*, London (May 10, 1984); **Orthopedic surgeons**: *Guardian*, Manchester, UK (Dec. 18, 1991). **84: Baltrip**: *Sunday Morning*, TV program, Augusta, Georgia (Jun. 5, 1988); **Black Sea firefighters**: *Chronicle*, Houston, Texas (Jun. 5, 1974); **Lingenfeld f/fighters**: *Daily Telegraph*, London (May 10, 1986); **San Antonio f/fighters**: *Times Reporter*, San Antonio (Mar. 23, 1991); **Sedgewick**: S. Pile: *The Book of Heroic Failures* (Kegan Paul, London, 1979). **85: Burgler**: *Niagara Falls Review* (Jul. 3, 1985); **Morris**: *Sunday Express*, London (Jan. 28, 1986); **St. Mungo**: *Glasgow Town & City Guide* (A.A., UK, 1988); **Shipman**: *Associated Press* (Sep. 10, 1985); **Firth**: *Daily Mirror*, London (Jul. 19, 1979); **Thornton**: *Sunday Mirror*, London (Jul. 18, 1982); **Delius**: *Daily Express*, London (Oct. 11, 1988). **86: Reyn-Bardt**: *Daily Telegraph*, London (Dec. 13, 14, 15, 1983). **87: Canny**: *Journal*, Lincoln, Nebraska (Dec. 28, 1987); **Fenwick**: *Daily Telegraph*, London (Dec. 24, 1988); *Express & Star*, Wolverhampton, UK (Dec. 24, 1988); **Wiezal**: *Daily Times*, Pawtucket Valley, Rhode Is. (Mar. 2, 1987); **Martin**: *Sun*, London (Sep.

6, 1988). **88–89: Son of Sam**: info supplied by M. A. Hoffman II. **90–91: Coincidences**: C. G. Jung: *Synchronicity* (Ark, UK, 1991); **Synchronicity**: ed. L. Picknett: *The Encyclopedia of the Paranormal* (Macmillan, UK, 1990).

Chapter 4: Eye of the Beholder
93: Hopkins: *The Unexplained* (Orbis, UK, 1981). **94: Christie**: *Shropshire Star*, UK (Apr. 2, 1989); *The Unexplained* (Orbis, UK, 1981); **Serres**: *Daily Telegraph*, London (Aug. 4, 1975). **95: Stillsmoking**: *Daily Mail*, London (Feb. 22, 1979); **Tobacca**: *Los Angeles Times* (Sep. 7, 1988); **Carrots**: *London Evening News* (Jun. 27, 1979); **Chickens**: *Daily Telegraph*, London (Jul. 25, 1979); **Eels**: *Scunthorpe Evening News*, UK (Sep. 24, 1982); **Drake**: *Daily Telegraph*, London (May 29, 1985); Christie, Manson & Woods Ltd. sale catalog, London (May 28, 1985); **Reverse hills**: info supplied by C. Cooper. **96: Kern**: *Philadelphia Inquirer* (May 2, 1983). **97: D-Day**: *The Unexplained* (Orbis, UK, 1981); *Strange Stories, Amazing Facts* (Reader's Digest, 1976). **98: Bacup plank**: *Sun*, London (Nov. 25, 1988). **99: Forest spirit**: info supplied by D. Efthyvoulos. **100: Griffiths & Wright**: *The Unexplained* (Orbis, UK, 1981). **102: Doyle**: A. Conan Doyle: *The Coming of the Fairies* (Hodder & Stoughton, UK, 1922). **103: Fairy photos**: ed. L. Picknett: *Encyclopaedia of the Paranormal* (Macmillan, UK, 1990); info supplied by J. Cooper; K.I. Jones: *Conan Doyle & the Spirits* (Aquarian Press, UK, 1989). **104–106: Images on eyes**: *Fortean Times*, UK (Spring 1987). **106: Mrs. Smith**: *York Gazette & Daily*, Pennsylvania (Apr. 4, 1940); **Comet eggs**: info supplied by P. Sieveking; **Christ Church**: *Daily Express*, London (Jul. 17, 1923). **106–107: Vaughan**: *Notes & Queries*, UK (Feb. 8, 1902). **107: Death's heads**: C. Fort: *Wild Talents* (C. H. Kendall, 1932); **Turin shroud**: I. Wilson: *The Turin Shroud* (Doubleday, 1978); *Strange Stories, Amazing Facts*; *The Unexplained* (Orbis, UK, 1980). **108: Superstitions**: E. & M. A. Radford, ed. & revised by C. Hole: *Encyclopaedia of Superstitions* (Hutchinson, UK, 1961); Zolar: *Encyclopedia of Omens, Signs & Superstitions* (Simon & Schuster, 1989). **110–111: Tutankhamen**: A. C. Brackman: *The Search for the Gold of Tutankhamen* (R. Hale, UK, 1978); *The Unexplained* (Orbis, UK 1981); T. Howing: *Tutankhamen — The Untold Story* (Hamish Hamilton, UK, 1979); C. Wilson: *Encyclopaedia of Unsolved Mysteries* (Harrap, UK, 1987). **111: Atreus**: R. Graves: *The Greek Myths* (Penguin, 1955); *Encyclopaedia Britannica* (1983). **112: Pharaoh's curse**: *The Times*, London (Dec. 27, 1991); *New Scientist*, UK (Jan. 11, 1992); **Cravens**: *Daily Telegraph*, London (Aug. 31, 1990); **Grimaldis**: J. Robinson: *Rainier & Grace* (Simon & Schuster, 1989); **Guinness**: *Sunday Times*, London (Sep. 3, 1978); **Squires**: *Western Morning News*, UK (Apr. 27, 1985). **113: Celtic heads**: *Incredible Phenomena* (Orbis, UK, 1984); G. J. McEwan: *Mystery Animals of Britain and Ireland* (Robert Hale, UK, 1986); **D362**: info supplied by P. Sieveking.

Chapter 5: Living Legends
114: Tendaevs: *National Enquirer* (Apr. 16, 1991); *Fortean Times*, UK (Sep. 1991). **116: Suvorov**: *Soviet Weekly* (Jun. 21, 1990); *Fortean Times*, UK (Sep. 1991). **Abbot**: J. Michell & R. Rickard: *Phenomena* (Thames & Hudson, UK, 1983); **Gaiduchenka**: *People*, South Africa (Apr. 16, 1991); **Magnet Annie**: *Phenomena*. **117: Mrs. Priestman**: *Sunday Express*, London (Jan. 20, 1985); **Ivanov**: *Sunday Express*, London (May 20, 1984); **Xue Dibo**: *Fortean Times*, UK (Summer 1988). **118: Mummified figure**: J. Brandon: *Weird America* (E. P. Dutton, 1978); **Little people**: *Encyclopaedia of Religion* (Macmillan, UK, 1987); **Neolithic graves**: *Folklore* Vol. 43 (London, 1932); **Gog & Magog**: *Man, Myth & Magic* (Purnell, UK, 1970–71); **Tom Thumb**: *Into the Unknown* (Reader's Digest, 1981). **119: Pygmies**: C. M. Turnbull: *Peoples of Africa* (Holt, Rinehart & Winston, 1965); **Unhappy giants**: *Into the Unknown*. **120: Leprechaun**: W. B. Yeats: *The Celtic Twilight* (Bullen, UK, 1902); **Lehane**: B. Lehane: *Companion Guide to Ireland* (Collins, UK, 1973); **Crock of gold**: E. E. Evans: *Irish Folk Ways* (Routledge, UK, 1979); **Gulliver**: *Encyclopaedia Britannica* (1983); **Finn MacCool**: P. Haining: *The Leprechaun's Kingdom* (Souvenir Press, UK, 1979). **121: Kuvera**: *Encyclopaedia Britannica* (1983); **David & Goliath**: *Man, Myth & Magic*. **122: Falashas**: *The Times*, London (Jan. 4 & 5, 1985; Mar. 23, 1985; May 27, 28, & 29, 1991); **Lost tribes, Adair, Madoc**: *Mysteries of the Ancient Americas* (Reader's Digest, 1986). **124: Melungeons**: *Mysteries of the Ancient Americas*; *Weird America*. **125: Roanoke**: F. Edwards: *Strange People* (Pan Books, UK, 1966); *Strange Stories, Amazing Facts* (Reader's Digest, 1976). **Ishi**: T. Kroeker: *Ishi in Two Worlds* (Univ. of Calif. Press, 1961)

Chapter 6: Too Good To Be True
128: Whale tumors: R. Dale: *The Tumour in the Whale* (Duckworth, UK, 1978). **128–130: Severed fingers**: J. H. Brunvand: *The Choking Doberman* (Penguin, 1984). **129: Angels of Mons**: *The Unexplained* (Orbis, UK, 1981). **Machen**: "The Bowmen": *Evening News*, London (Sep. 29, 1914). **130: Severed hand**: *La Nouvelle Fabrique des Excellents Traits de Vérité* (France, 1579). **132–135: Aug. 21, 1915**: *The Unexplained* (Orbis, UK, 1981); Paul Begg: *Into Thin Air* (David & Charles, UK, 1979); *The Final Report of the Dardanelles Commission* (UK, 1965). **135: Brain diseases**: O. Sacks: *The Man Who Mistook His Wife for a Hat* (Fontana, UK, 1985); **Phantom vicar**: *The Unexplained* (Orbis, UK, 1981). **135–136: Gauquelins**: *Skeptical Inquirer* (Winter 1979–80 & Summer 1980); *Cosmic Connections* (Time-Life, UK, 1988). **137: Dighton Rock**: G. Mallery: *Picture-Writing of the American Indians* (1893); J. Michell: *Megalithomania* (Thames & Hudson, UK, 1982). **138: Slabbert**: *Daily Mirror*, London (Oct. 13, 1987); **Prague**: *News of the World*, London (Feb. 15, 1987). **139: Aspaugh**: *The Unexplained*, (Orbis, UK, 1981).

INDEX

Page numbers in **bold** type refer to illustrations and captions.

PHOTOGRAPHIC SOURCES

AA Picture Library: 85; **Ancient Art & Architecture Collection**: 67 (B. Wilson); **ASAP**, Jerusalem: 121tl (Israel Museum), 123 (A. Auerbach); **Associated Press**: 28b, 88b (H. Goldenberg); **John Beckett**: 102; **The Booth Museum of Natural History, Brighton/courtesy of David Leon**: 60-1; Reproduced by courtesy of the Trustees of the **British Museum**: 95; **Bruce Coleman Ltd.**: 24, 56t (G. Zeisler), b (G. Langsbury), 57b (L. Lee Rue), 58t (S.J. Krasemann), 66t (P.A. Hinchcliffe), b (H. Reinhard), 70b (F. Lanting); **C.M. Dixon**: 118br, 121tr, cr; **Thomas S. England**: 74t; **Mary Evans Picture Library**: 21(inset), 27, 37t, 38, 60t, 68l (Harry Price Collection, University of London), 69t&b (Harry Price Collection, University of London), 89t, 96, 100-1, 119br, 120r, 129, 136; **Fortean Picture Library**: 25, 36, 42 & 43t (Dr.J.T. Richards), 46 (Shin-Ichro Namiki), 47b (F.C. Taylor), 48t, b (G.T. Meaden), 57t (A. Barker), 68r, 70-1 background, 70t (A. Shiels), 98tl (Borderline), 99br (P. Broadhurst), 106b (S. Lansing), 107t (Dr. E.R. Gruber), 113t (P. Screeton); **Colin Godman**: 107b (SIPA); **Ronald Grant Archive**: 94t; **Robert Harding Picture Library**: 110 (D.W. Hamilton), 111, 118l, 121b (T. Wood); **Hulton-Deutsch Collection**: 103, 112, 118tr; **Hutchison Library**: 119t (Davey), bc (Errington); **Image Bank**: 34-5 (C. Feulner); **Images Colour Library**/Charles Walker Collection: 62, 89b, 98b, 99bl; **Imperial War Museum**: 50-1; **Andy King**, San Antonio: 84b; **Kobal Collection**: 33, 75t, 130t; **Natural History Picture Agency**: 71t (M. Bain); **Peter Newark's**

Pictures: 84t; **Pan MacMillan**: 135b; **Popperfoto**: 113b, 125b; **Press Association**: 86; **Public Record Office**, Crown Copyright: 97; **Retna Pictures Ltd.**: 130b (C. Lipson); **Rex Features**: 28t, 47t (Today), 94b, 107b (SIPA); **Royal Norfolk Regimental Museum**: 132-3; **Science Photo Library**: 43b (X-Ray Astronomy Group, Leicester University), 77t (St. Bartholomew's Hospital); **Frank Spooner Pictures/Gamma**: 63 (Rivere), 71b (Mansi/LSN), 74b (McKierman/Liaison), 76 (Kermani/Liaison), 99t (D. Efthyvoulos), 117b (Novosti), 124 (P. Bartholomew/Liaison), 131; **Stanford University Archives**: 39; **Syndication International**: 77b, 98tr, 117t; **Tass**: 116; **Telegraph Colour Library**: cover (inset); **Colin Twissell**: 37b.

b - bottom; c - center; t - top;
r - right; l - left.

Efforts have been made to contact the holder of the copyright for each picture. In several cases these have been untraceable, for which we offer our apologies.